The Quest for the Dhirniy

By Ralph Battaline

ISBN: 9781961563308

DEDICATION

This book is dedicated to those whose minds have walked the surfaces of countless worlds, and have shaken the hands of many races.

ACKNOWLEDGMENT

I would like to thank my Beautiful Eternal Companion, Cati, and my three wonderful children, Isaac, Elara, and Theresa.

They supported me as I worked to complete this story.

I also thank Kimberly for the inspiring ideas that enabled me to continue with the story when my ideas ran dry.

Special thanks go out to the publishing staff, who worked very hard to make a vision come to life.

Most of all, I give thanks to my Father in Heaven for enabling me to see the story as I wrote it. This was truly a gift. The pen practically moved on its own.

Contents

About the Author

I was born in Connecticut as the youngest and oddest of five children. I grew up with a strong fascination with the universe. I became a science and mathematics instructor after receiving my Earth Science and Astronomy degree. I would like to believe that I positively influenced the minds of literally thousands of students. I was also a planetarium director for different museums and schools.

I bought my first telescope as a teenager, whereas others my age were thinking about their first car. I hope you remember me saying I was the oddest of five children. With that telescope, my mind traveled through the universe.

Allowing myself to return to Earth, I was sealed for time and all eternity to the most beautiful and wonderful woman a man could ever hope to share eternity with. She has filled my life with indescribable happiness, which I will never be able to repay. We have been blessed to be the parents of three wonderful children. We love watching them grow as they fight with each other continually.

I will stand in awe forever at the majestic workings of our Creator's hands, and the blessings He bestows upon us daily.

Ralph Battaline

Page Blank Intentionally

PREFACE

Written initially as a computer game, I await a talented and ambitious team of game developers to take this work of interest from paper to the computer and television screen.

My drawback is, like many, great ideas but not enough money. I hope I will come to know that a special group of creators are willing to show faith and accept payment as the game succeeds.

I am willing to give that special group 30% of what the game brings in.

It may seem that I'm living in a dream state, but it's all I have for the present.

I hope you will enjoy "The Quest for the Dhirniy" story. If you are one of those talented I seek, I know you will see what I have seen. Please make contact with me. This much I know: the game is a winner waiting to happen.

R. Battaline

astroman1226@ gmail.com

INTRODUCTION

The Quest for the Dhirniy (Dear-Nye) takes place in the year 2052. It is a time when space travel is of great interest in the lives of mankind. Traveling at the speed of light is the goal many scientists aim to reach. Richard Gaber, an ambitious scientist working for TransQuest, a company specializing in future engine design, has secretly achieved speed of light travel. Not believed by his colleagues, Mr. Gaber confides in one man to share his findings: a quiet but inquisitive individual named Lee Travis, a Captain from his days in the military.

You are to take the role of Captain Lee Travis and embark on a journey that becomes a mission far greater and more dangerous than what could ever be spoken in words. You will face life and death challenges at every point along your journey as you unravel what could be a new chapter for mankind should you succeed, or your destruction should you fail. Your mission is to unite two worlds in a friendship never before understood.

Make the right choices, and you will follow an ancient trail that will take you far beyond where any human has ever traveled. Make the wrong choices, and you are no more.

PART 1
THE INVITATION

CHAPTER 1
THE TRAVIS RESIDENCE

Home at last! I had no idea returning to a normal routine would be so difficult. The past couple of work days felt as though they would never end. What used to be an exciting but questionable goal at work has vanished away, replaced by knowledge and experience. I just had an adventure that I will never forget. My entire life has been changed forever, and I will never see things the same again.

Oh, please forgive me for rambling on. Allow me to introduce myself. My name is Captain Lee Travis, retired Air Force pilot. Trust me when I say that you can never know how events in your life will unfold from one day to the next. Take me, for example. If somebody told me that I would meet a race of people from……., well, perhaps I should start at the beginning.

I work for a company called TransQuest, where I find my job to be very interesting. Personally, after being honorably discharged from the Air Force, I've wondered now and again if my current labors would ever amount to anything. You see, I build engines for spacecraft. Now I know that doesn't seem odd when you hear it, but these engines are special. Building engines for spacecraft is one thing, but to build one that can achieve the speed of light! That might be asking a bit much. Think about it! The speed of light, the universal speed limit, the unreachable velocity for man-made objects.

I've been working with TransQuest Corporation for years, and I still have difficulty wrapping my head around the idea. It's not like traveling 70 mph on a highway. We are talking about a speed of 670,617,741 mph (186,282.7 mps). At such relativistic speeds, rather hard-to-explain conditions occur. Mass grows towards infinity, length decreases in the direction of motion, and time slows down. These are not thoughts that one wishes to think about when trying to fall asleep. Then again, I'm so tired right now; I don't think there is much that will keep me up for much longer. I just want to take a load off my feet and rest on the couch for a little while.

AHH! That's much better. I mustn't fall asleep though; I still have a few things I need to do. Still, I can't believe how far the company has progressed in its attempts to build such a phenomenal engine. Not too long ago, the company's best speed was achieved; a velocity of 5.52% the speed of light. Five and a half percent out of a hundred may not seem like very much, but it took a great deal of effort to achieve it. If it wasn't for the determination and unconquerable spirit of my friend and colleague, Mr. Richard Gaber, the company would still be floundering about with no real belief that such an engine could actually be built. I'll never forget how it all happened. Oh, and by the way, I must apologize upfront. I have a habit of thinking aloud, often with considerable detail. I'm sure I'll often jump from past to present tense while explaining my incredible adventure. But I feel that if you are to understand the events that took place, you must feel and see them in your mind, as I did with my eyes.

With that being said, it started about a month ago. I was sitting on this couch, exhausted, just as I am now. I noticed that I had a transmission waiting to be seen and heard. I reached over and activated my communications device, and found myself looking at the face of Mr. Richard Gaber. At the time, Mr. Gaber was a colleague his fellow workers judged. They didn't believe he had all of his oars in the water. Mr. Gaber was a man who kept to himself at work because of how he knew his colleagues felt about him. His colleagues often ridiculed him, thinking of him as very impractical. These feelings came about because of a claim that he made. Rich announced that he was coming close to achieving the company's goal of creating an engine capable of traveling near the speed of light. Most of the company's personnel felt that the goal was unreachable, and that Mr. Gaber was a dreamer. The following is what I saw and heard on my communications device the night my adventure began.

"Hello, Captain Travis: this is Richard Gaber. I'm sorry to bother you at home. I debated as to whether I should contact you or not. Despite the reputation others have given me, I have decided to do so, and with good reasons. I would have spoken with you directly at work, but I felt a private conversation would be best. I have dealt with enough ridicule already to last me the rest of my life. Perhaps you feel as the others do, that I am a dreamer with an overactive imagination. I hope this is not so.

I am calling you because you are more open-minded than the rest of our colleagues. Conversations with you have led me to this conclusion. I, therefore, feel you are the only person who will take me even half-seriously. I know that I come off sounding completely obsessed with my beliefs on the possibilities of building a spacecraft capable of carrying mankind beyond the solar system. Perhaps I have been overly involved with my ideas and research concerning the matter, but I know my dedication will pay off in the end.

As you are aware, I have been working for some time to increase the velocity capabilities of our spacecraft. The company's best so far is an arrival time of just over four days to Pluto. This results in a velocity of just under 5.6% of the speed of light. If I'm not mistaken, it is the craft you and a small group of others have mainly been working on. I hope my suggestions were of value to the group."

"Lee, what would you say to an arrival time of twelve minutes or less to Pluto?"

"Have I sparked enough of your interest, or have I made you think me totally insane?"

"I've done it, Captain Travis. Traveling at very near the speed of light is now possible. I am living proof of it, and I wish to prove it to you. If you are interested, please convince yourself to believe in what I have said. I ask that you meet me at TransQuest no later than 8 p.m. this evening. It is imperative that you come on time, Captain Travis. When you arrive, check your mailbox for instructions on how to locate me. I will not be in my office. I hope to see you tonight at eight, Captain Travis."

After the message ended, I stared at my communications device with my mouth slightly open in astonishment. Is it possible that Mr. Gaber achieved what he believed in so strongly? I looked at the clock on my living room wall, which displayed a time of 7:21 p.m. I had to move quickly if I was going to meet Mr. Gaber by 8 p.m., but would it be worth it? I guess there was only one way to find out. As tired as I was, I got up from the couch and headed out the door, hoping this would not turn out to be a disappointment.

I went to my vehicle and placed my hand on the door to open it. My hand was scanned and the door opened to allow me entrance. Once inside the vehicle, I saw a list of pre-programmed destinations displayed on a little screen, much like a graphing calculator. Each destination was represented by a number coordinate with the name of each place written next to the coordinate. I found the listed coordinate for TransQuest, 007326, and input the coordinate for an automatic ride to my place of employment. Eight minutes later, I found myself back at TransQuest.

Chapter 2
In Search of Mr. Gaber

Having arrived, I exited my vehicle and approached the building I had just left not more than an hour ago. Upon reaching the main entrance, I was greeted by the company computer system that asked me to input my identification number into the keypad alongside the door. I input 75192. I was then told to look into the retina scanning device above the keypad. After scanning my retinas, the computer sweetly said, "Good evening, Captain Travis," and the main entrance door opened automatically.

Once inside the building, I remembered the instructions to check my mailbox. While on my way, my attention was drawn to the main reception desk. I took a small detour over to the desk where I found a book titled "Our Sun's Family;" with a message on it from Mr. Gaber to me. Mr. Gaber's message said the following:

Captain Travis, please keep this book in your possession. I am certain that it will be of great value to you.

I opened the book to find it filled with information about the planets and satellites of our solar system. I couldn't see how a book on planetary astronomy would ever be of value to me. Nevertheless, I picked up the book and continued on my way to my mailbox. When I got there, I found two items within it; a torn fragment of paper with the number 226912 written on it, and a handheld communications device with a screen within it. I saw that the device had an active file waiting to be displayed. The following message was heard with Mr. Gaber's face on the device's screen.

"Captain Travis, I am glad that you decided to come. I know you are on time because, at 8:01 p.m., this device was programmed to erase the message you are now listening to. Therefore, this message would not be available if you were late. As I told you earlier, I am not in my office. Please look for me now in the room whose number equals Pluto's rotational velocity minus Luna's rotational velocity plus an additional 0.09. To find this information, look for a book I left for you on the secretary's desk if you don't have it already. Remember, Captain Travis, do not be late. See you soon."

After hearing this message, I looked up at the clock on the wall and saw it display 7:39 p.m. I thought the race was over. I now had 21 minutes left to find Mr. Gaber. I immediately opened the book I took off the reception desk to look up the rotational velocities of Pluto and Luna. I read that Pluto rotates at a velocity of 30.26 mph, and Luna at 10.35 mph. I didn't realize that those two bodies rotated so slowly. Subtracting Luna's rotational velocity from Pluto's and then adding 0.09 gave me an answer of 20. Mr. Gaber was, therefore, in room 20 of this building.

Looking around, I saw that all room numbers were in the low 100s. I needed to get downstairs. I found a staircase to get to the basement. Once there, I followed the hallway down, noticing the room numbers increase as I did. The odd-numbered rooms were on my left, which was the inner wall, and the even-numbered rooms were on my right, which was the outer wall. I reached the end of the hallway and found room 18. I then turned the corner and was shocked to see room 22 starting the new hallway. I looked back at the left side of the wall to see that the previous hallway ended with room 19, and that the new hallway started with room 21.

Where was room number 20? I couldn't believe I had worked here for as long as I had, and never realized there wasn't a room 20. My job duties never really took me down to the lower levels very much, but still. Curiosity had now gotten the better of me, so I began a search for room 20 with very little time left to succeed. I eventually located it by entering room 18, which I stood closest to. I later found out that by entering room 22, I would have had the same success.

I learned that room 20 was a room within a room. The way the building was designed, the entire corner of the building was a room that could only be accessed by entering one of its adjacent rooms; in this case, rooms 18 or 22. When I entered room 18, I found nothing out of the ordinary save a garage-looking door on the room's left wall toward the back. As I approached the odd door, I noticed that it had a keypad off to its right side, much like the keypad at the main entrance to the building. Not knowing what code to input, I naturally tried my own identification number. To my astonishment, the door began to lift open.

Once inside the unlabeled room, I stood in a very large area that occupied the entire corner of the basement floor. Inside the room, I saw an incredible-looking spacecraft, unlike any the TransQuest Corporation had been constructing. I had found the secret location where Mr. Gaber conducted his research, away from the ridicule of his fellow workers. Unfortunately, I could not find any sign of Mr. Gaber.

Shortly after entering the spacious room, the garage-looking door began to close. About five seconds after the door to room 20 had completely closed, a holographic image of Mr. Gaber appeared and spoke the following:

"Captain Travis, I am glad you succeeded in finding this room, and that you made it on time. Like the programmed message in your mailbox, this holographic program was also set to erase at 8:01 p.m. I have kept this room well protected from others in the company, and have received approval from our company president to privately use this space to work on my ideas. I have, therefore, programmed only two identification badge numbers to open this room; mine and yours. I'm sorry that I did not tell you about my endeavors until now, but it was necessary to keep my work undisturbed until I could properly discuss it. I share this information with you now so that you may experience and witness something I dare not try to put into words. I trust I have piqued your interest, so let's get started. Please forgive me for not being here in person as I originally planned. I was needed to attend to an important personal matter which relates to everything I need to explain to you. Please be patient with me, my friend. All of this will make sense shortly.

First, I would like to say that I have always believed in this company's goals, of which you are quite well aware. From my first day of work at TransQuest over seven years ago, I have worked on constructing a spacecraft that could travel near the speed of light. As difficult as it was, I put together different working methods with the rest of the team. When the team worked together, there were promising moments. After a great deal of effort, I finally hit the jackpot. I would come to this room night after night when everyone else had left for the evening. It was then that I would work on my ideas. I would build one engine after another and test it in a spacecraft. I would take the craft out of the hanger door in the ceiling and give it a test drive. When I returned, I would make improvements based on the previous flight.

At one point, I had built an engine capable of traveling about 87.2% the speed of light. This velocity enabled me to travel beyond the solar system's borders, and to explore a number of others. In doing so, I came across a race of beings that call themselves the Dhirniy. (Dear-nye) I found them to be very benevolent, welcoming, and quite intelligent. All of this took place about three years ago. I would have liked to share my discovery with you then, but I knew the moment was not right. Please take no offense to this, for none was ever intended.

I, therefore, built relations with the Dhirniy for the last three years. I have learned much about this race of people, and they have learned a great deal about us as well. They helped me understand how to improve our engines so that they would travel near the speed of light. During my last test flight, I achieved a speed of 99.48% the speed of light. I have enjoyed getting to know this wonderful race of people, and I believe they have also enjoyed my visits as well.

In learning about us, they have seen that we are still a child race. Even so, they welcome our race to their world, but only those who can display the higher qualities that inherently exist in mankind. The negative and careless ways of man, such as hatred, greed, and so many other weaknesses, are not dealt with anymore by the Dhirniy. Although far from perfect, they have overcome their childlike period. They fear that if others of their world were exposed to it again, it could only bring about eventual ruination to their world. I spoke to the Dhirniy President, Axnard Jirroch, and told him about you and our relationship. I informed President Jirroch that you are a man of higher principles. You have, therefore, been invited to their world called Marathrax, should you wish to come.

To prove the qualities of an individual, the Dhirniy have put together a challenge in which compassion, endurance, courage, and intelligence are tested. The challenge is that the individual finds a message written by the Dhirniy, instructing the individual how to locate the Dhirniy home world. The message is broken into twelve pieces, which, when put together, will reveal the course to Marathrax. The pieces of the message are placed on different planets and satellites within our solar system. The locations where the pieces of the message reside are listed within the computer on board the ship you see before you. If the individual can find the Dhirniy home world, and a few extra tasks along the way, they will have exhibited the higher qualities in all mankind. The Dhirniy believe that all intelligent creatures possess unconquerable goodness.

Captain Travis, I do not need to tell you that this mission will be quite challenging and extremely dangerous. I will understand if you feel the task is too demanding and wish to decline. I will not lose the respect and friendship I have come to feel toward you. Therefore, if your interest and curiosity are still piqued, please follow me at this time, and I will show you the ship I have worked on. I hope you will be impressed by what you see."

After Mr. Gaber's hologram stopped talking, it waited to see what I would do. I thought my mouth was hanging open quite a bit after the initial message from Mr. Gaber. Trust me when I say that it was nothing compared to how much it was hanging open now. This was amazing! There in front of me was a holographic image of Mr. Gaber, inviting me to follow him into a remarkable-looking vessel. He stood there waiting to see if I would follow him or turn and walk away. I tried to ask Mr. Gaber's hologram questions, but it was not programmed to respond to them. All the same, I wasn't about to walk away after all that I just heard. It was more than obvious that Mr. Gaber was telling me the truth. Therefore, I walked closer to the holographic image of Mr. Gaber, which then turned toward the ship and walked inside. A few moments later, I, too, entered the craft.

Chapter 3
Aboard the Lotania

Moments after entering the ship on its starboard side, Mr. Gaber's hologram turned toward me and spoke again.

"Welcome aboard the Lotania, Captain Travis; this will be your ship."

My mouth hung open a third time, and my eyes widened as I listened in amazement. I felt like my mouth opened far enough that my bottom lip was capable of sweeping the floor. After a small pause, which he knew would be necessary, Mr. Gaber continued to speak.

"Yes, you heard me correctly. This ship will be yours to search for the instructions to the Dhirniy home world. There is no other craft like it, except the duplicate craft I built, which took me to Marathrax initially. Please allow me to familiarize you with the ship so you will be acquainted with its operations and cargo.

Let us begin with the items on board the craft that will aid you with your mission. First, we have the Land Rover, which you will need to use on some of the places you will visit. The use of the Land Rover is quite simple. Drive the vehicle as you would an old car. It has an accelerator and a brake pedal. The rover also has a robotic arm attached to its roof, as you can see. To operate the arm, just use the rover's joystick control to move the arm in the direction you wish it to go. To open and close the jaws on the robotic arm, you must squeeze and release the joystick controller.

Now, if you look across from the Lotania's main entrance, you can see the ship's cabinet containing various space suits, footwear, food, and a backpack containing tools and other items you will need on your mission. The backpack contains a flashlight, a handheld laser, a cable gun with a strong one-hundred-foot metal cable capable of retracting, and various other items.

You will have three different suit and boot types. One suit is for normal space conditions, one is for extreme heat and pressure, and the last is for underwater exploration. Yes, you heard me correctly. My friend, I can only imagine your questions at this time, but we must continue. All of your questions will be answered shortly.

The three types of footwear you have are for normal space conditions, weighted boots for low gravity conditions, and cleated boots, which are weighted for icy conditions. Please be aware of the conditions of all the places you will visit. This way, you can dress correctly for a safe excursion. I cannot stress this enough. To help you with the conditions of the planets and satellites, look over the book you picked up from the secretary's desk. Did I not say that the book would be of great value to you?

On the other side of the ship again, next to the main door, is the ship's toolbox. The normal tools you would expect to find in any toolbox are in it. In addition to your common tools are important instruments like an ice pick, heavy-duty cutters, and tire chains for the Land Rover. As you travel, expect to find additional items to aid you in your quest.

Let us now draw our attention to the Lotania's control panel. Won't you please have a seat, Captain Travis?"

I watched as Richard's image gestured me to a chair. I then sat down at the ship's control panel at the request of Mr. Gaber. He then continued speaking.

"From your seat, you should be able to reach all areas of the board. Please recognize that all of the controls are labeled for you. Starting on the left side, we see the button for the hanger door to this room, room 20, that is. Pushing the button opens and closes the hanger door located in the ceiling by remote control."

Although Mr. Gaber's image began on the left side of the panel, I noticed that it didn't mention a number located on the far left area of the panel. The number was 226912, the same number I found in my mailbox when I arrived at TransQuest. Unable to ask questions, I continued listening to Mr. Gaber, who continued describing all the functions of the spacecraft. The Lotania was equipped with many incredible functions, which included shields, weapons if needed, automatic and manual control, cables to stabilize the craft in low gravity conditions, and so much more. I listened intently to all that Mr. Gaber's hologram said. I especially focused when he spoke about inputting coordinates and using the Lotania's weapons and radar systems. Richard went on to say the following about these three areas:

"Next over is the computer's data monitor. On this monitor will be displayed the very important list of coordinates that correspond to the places in the solar system that have the items you seek. When you are seated and the Lotania's door is closed, SAROS, the Lotania's computer, will ask you to input your destination coordinate. When the ship is on automatic, you only need to input a coordinate into SAROS. When entered, press the engage button, and the Lotania will take you to the location matching the coordinate you input."

"Captain Travis, what I tell you now, I say with all seriousness. Please take a moment to read the warning written under the last coordinate; it is very important."

Mr. Gaber paused a few moments, giving me time to read the warning. Rich continued by saying,

"Please take this warning very seriously, Captain Travis."

The written warning read as follows:

Do not input any other seven-digit coordinate than those listed above; the destination is uncertain.

Rich then continued.

"Next are the Lotania's weapons. The ship comes with lasers and torpedoes. Although the Dhirniy are a peaceful race and do not support violence to solve any problem, they understand the need for self-preservation. I say this because the Dhirniy are having trouble with a neighboring race called the Allorayans, a race of beings who feel the need to conquer the Dhirniy. The weapons are a precaution should you be mistaken for the enemy. You have forward and aft laser and torpedo buttons. I hope that you will never have to use them."

Richard finalized his explanations of the ship with the following:

"My friend, I know that I have given you a lot of information, but please stay with me. Please focus your attention here; this is your ship's radar screen. Use it to see your surroundings, especially for Allorayan ships. The Allorayan ships will appear as yellow circular dots on the radar screen. By touching one of the dots, SAROS will display all known information on the ship. Your alert lights are the red and yellow lights to the right of the radar screen. If the yellow alert lights up, it means that possible danger is near but not on the radar as of yet. It would be best to raise the Lotania's shields at this time. If the red alert lights up, danger is definitely present and on the radar screen. You better have your shields up by now.

Below the radar screen is the Lotania's most lethal weapon. It is called the Roll, Pitch, and Yaw, RPY for short. This button, when pressed, will send the Lotania into a frenzy of axial spins. While spinning apparently out of control, the Lotania will fire a one-time load of lasers and torpedoes. This button should only be used if the Lotania is hopelessly outnumbered while surrounded by enemy vessels. Be warned, Captain Travis, only the vessels seen within the radar's inner circle are reachable by the RPY firing action. Once this ship function is used, it cannot be used again. Using the lasers and torpedoes associated with the RPY weapon will not affect the lasers and torpedoes of the Lotania. The RPY weapon has its own reserve separate from the rest of the ship.

Well, my friend, I have told you all that I can, and I'm aware that I talked your ear off, for which I do apologize. Please be assured though, that it was all very necessary. The rest is up to you. All you need to do now is begin your mission or turn around and go home. If you accept the mission, please be safe, my friend, and I will see you soon. If you leave, then I thank you for coming down to check everything out, at least. With either choice, I ask that you keep what you have seen and heard between us. You should know that if you leave the entire building after having entered the Lotania, all of the Lotania's programming will be erased. I hope to see you soon, Captain Travis. Take Care."

When Mr. Gaber's hologram finished talking, it faded away. Needless to say, my head was spinning after all that I heard. I was just given a tour of a spacecraft capable of traveling near the speed of light. I was just invited to travel to another world and meet an entirely different race of beings. Incredible! The mission was undoubtedly going to be extremely dangerous, but how could I possibly walk away from such an opportunity? I would kick myself for the rest of my life. Therefore, my decision to move forward with the mission was made.

Still seated at the control panel, I looked over the controls once again. The ship indicated that it was on automatic control, which is where I fully intended upon leaving it. I had no intentions of wrestling with the Lotania until I became comfortable with its maneuvers.

It was time to begin a journey of a lifetime, which life I hoped would not be cut short. I closed the Lotania's main door and opened the door in the ceiling so the Lotania could safely launch. The ceiling door rumbled open for about 30 seconds while the Lotania's main door closed. Once all the doors were positioned where they needed to be, I was briefly startled by the ship's computer, SAROS, which said:

"Please input your destination coordinate."

I looked over the coordinates available to me and decided to take them in order. I, therefore, input the coordinates for the planet Mercury, which read 0512374. Once placed within SAROS, I hit the engage button.

My blood ran cold when the vessel lifted off the ground. *Do I have any idea what I'm doing?* I thought. It was a bit too late to ask that now. The ship was supposedly on its way to Mercury, and so was I.

CHAPTER 4
THE MERCURIAN QUIZ

I was soon in complete and indescribable awe as I looked upon a sky filled with stars more richly than I had ever seen before. I was above the Earth's atmosphere, heading toward a rather bright star-like object I assumed to be Mercury. I did not need to assume long. The speed at which the Lotania traveled enabled me to see that the apparently approaching object was a spherical celestial body. It was, indeed, Mercury. Despite the amazing view of Mercury, the brilliance of the stars continued to be breathtaking. Never have I seen such overwhelming beauty and clarity when gazing into space. My attention was quickly routed back to the ship when, shortly after the journey to Mercury began, SAROS, the ship's computer, spoke the following:

"Your destination is Mercury, the closest planet to the Sun. Its average distance from the Sun is 35,983,094 miles. Mercury's diameter measures 3,032.9 miles. The length of day is 58.81 Earth days. The orbital period is 87.97 Earth days. The average surface temperature measures 332.0 °F. Its atmospheric gases are essentially a vacuum. The surface gravity measures 37.8 % of Earth's. For other important information, please read the planetary book."

"This is truly unbelievable! I can't believe this is really happening," I thought.

The entire trip to Mercury took about four minutes. As I traveled, I watched Mercury appear to grow larger and larger as the Lotania approached it. I was blown away as I watched the craft descend to the barren surface below me. I was about to land on Mercury! As far as I knew, I would be the first human to step foot on this alien landscape.

When the craft landed, SAROS said:

"Destination complete. Your position is the central region of the Caloris Basin, the largest basin on Mercury, measuring 960 miles in diameter. Please reference the planetary book to properly dress for the conditions of Mercury. Please be aware that your suit can only sustain you for ninety minutes due to the Sun's overwhelming radiation at our close proximity."

This was fantastic! The computer was talking about dressing properly to walk on the surface of Mercury as though it were telling me to wear a sweater on a chilly day for a walk in the park. Nevertheless, I immediately checked what the proper attire should be for a Mercurian excursion. According to the book, to safely walk the surface of Mercury, I would need a space suit that can handle extreme heat and pressure. The pressure on Mercury is extremely low, and the temperature is very high. Due to Mercury's low surface gravity, I would also need weighted boots, cleated or not. Knowing what to wear, I retrieved the proper clothing and gear from the cabinet and began dressing. I could only imagine what I must have looked like in that suit. There were no mirrors on the Lotania. When properly prepared, I opened the main door using the appropriate button on the ship's control panel.

I stood staring at the view before me. Although completely barren and devoid of life, it was beautiful to behold. After a few deep breaths that seemed to take hours, I exited the ship. I was on Mercury! Much like the Moon's, my foot was buried in a few inches of regolith. When I lifted my foot, I saw the first human footprint left behind. It was my footprint. I now knew what Neil Armstrong felt like when seeing his footprint on the Moon, only at a much grander level.

Focusing myself, I remembered to begin a ninety-minute countdown to make sure I'd make it back before lethal exposure could take effect. As I stood completely amazed on this alien surface, I saw the Sun in all its brilliance. Its apparent size was definitely larger than what is seen on Earth. It was extremely odd to see the Sun in a black sky. This, of course, was due to there being no atmosphere on Mercury.

As I continued looking around, I spotted what appeared oddly enough to be a machine off in the distance amidst a desolate, cratered landscape. Walking about unrestricted while trying to reach the machine, I found it was not as easy as it looked. This was due to the lower surface gravity, heavy cratering, and other rugged surface features of Mercury. Walking toward the machine first took me to the rim of a crater. Although small, I saw I would need to enter into it to avoid going around it. Walking around the crater would take more time than I had available.

The wall of the crater was steep enough that my grappling hook gun would be needed to safely descend. Securing the grappling hook to a sturdy boulder, I began my descent. Repelling to the crater floor took less time than I anticipated. My only hope now was that the gun would retract the hook as Mr. Gaber's holo image said it should. Thankfully, the hook returned without incident. Once on the bottom, I walked to the opposite wall of the crater. As I did so, I was impressed by the rebound center that formed when the crater was made. When I reached the opposite wall, I found it was not as steep as the side I had entered. I, therefore, climbed up the crater wall very carefully, avoiding a few large boulders here and there.

After exiting the crater, I resumed walking toward the machine. I noticed that I had sixty-three minutes left until possible danger would occur from exposure. About fifteen yards from the machine, I saw a partially buried sign in the Mercurian regolith. The only thing on the sign was the mysterious number 226912. "*What is it with this number?*" I thought. "*What is its significance?*" Looking over the sign front and back, I could find no explanation. With little time left, I had to get to the machine. I'd deal with the mysterious number later.

Upon reaching the machine, I found it covered with a thick layer of dust, which I removed as best I could. When fairly clean, I saw that the machine's surface was rectangular. In the upper right-hand corner, I saw a small square metal covering. A red button under a clear protective covering was in the lower right-hand corner. To the left of the rectangular compartment, there was an empty area that resembled a television screen that was turned off. Beneath the blank screen was another metal covering the length of the blank screen.

To get things underway, I saw that I needed to find a key that would start the machine. Examining the machine more closely, I located the key under the metal covering in the upper right-hand corner of the machine. I assumed the metal cover was for keeping out dust contamination. I could see nothing else special about it. Taking the key, I started the machine, which I was amazed still operated. I couldn't begin to guess how long the machine had been sitting there.

When the machine started, the metal covering beneath the blank screen revealed a keyboard similar to a computer. The area resembling the television screen above the keyboard lit up with words to be read. On the now-activated screen, I saw three questions, which read:

"What is the Sun's diameter?" "How far away is Mercury from the Sun?" and "What is the Sun's angular size as seen from Mercury?"

Alongside each question was a space to input an answer. I had no idea what the answers were to these questions. I was then very happy to remember that I brought along the solar system book. I just hoped the information was in there. Flipping through its pages, I found the chapter dealing with the Sun. All the answers I needed were there except the last one. Looking through the information in the book, I found nothing about the Sun's angular size, or any other object for that matter. I continued to search the book for anything that might lead to the answer.

My time was running out. I needed to get back aboard the Lotania in thirty-nine minutes. I almost gave up and began heading back to the ship when I saw a page of formulae at the end of the book. They were all listed to tell what they would solve. One of them was listed as angular size. Looking over the equation, I filled in my information into the formula. I calculated an answer which I hoped was correct and quickly entered all the answers to the questions into the machine.

I must have answered correctly. When done, the clear covering over the red button on the machine flipped open. In a near panic over my time concerns, I immediately pushed the red button without thinking. I can't say that my action was one of my more brilliant ones, but I didn't wish to face the conditions of Mercury should my time run out. If I was going to go, I'd rather it be with a possible bang.

When the button was pushed, a metal box rose on a small elevator out of the center of the Caloris Basin. In cautious amazement, I slowly went over to the box, picked it up, and examined it. Seeing nothing special about the box, I opened it slowly and found a fragment of paper within it. On the paper fragment was writing that I could not understand. It was written in a language I had never before seen. *This must be one of the puzzle pieces Mr. Gaber told me about,* I thought. I had found the first one and was still alive. Maybe I would see this Deernish home world after all, or however it's pronounced.

With renewed confidence, I carefully placed the fragment of paper with my gear and quickly returned to the Lotania. The countdown now read twenty-four minutes. It took me more time than that to get to the machine. I really needed to hurry. With no time to lose, I followed my tracks back to the crater and entered it again. I hastily climbed down the side of the crater, almost losing my balance a couple of times.

Moving with as much speed as possible with a suit on and being subjected to a gravity much lower than Earth's, I made my way to the other side of the crater. I removed my grappling hook gun, took aim, and fired at an outcropping rock beyond the crater rim. Thankfully, I hooked the rock with my first attempt. I activated the gun's retraction function to help me move more quickly up the crater wall. When I reached the top, I quickly detached the gun's hook and ran for the ship.

When I entered the craft, I closed its outer door. I saw that I had returned with 54 seconds left on the clock. Relieved to be back, I collapsed to the floor and rested for a few moments. I laid there for about three minutes, going over my unbelievable experience. Shaking myself free of my thoughts, I removed the space suit and boots and placed them back into their closet. Taking one more look at the fragment of paper, I saw characters that may as well have been hieroglyphs. I couldn't believe I was gazing upon a language from a world beyond my own. To keep it safe, I placed it in the ship's cabinet. I then returned to the chair behind the control console. This must have activated SAROS, who then said,

"Please input your destination coordinate."

I then looked up the coordinate for Venus, which was 1768124. I input it into the computer and pressed the engage button. Although not as bad as the first time, I still felt chills throughout my body as the ship ascended. I was on my way to Venus.

CHAPTER 5
VENUSIAN PERIL

Shortly after my trip to Venus began, I saw it coming up on me quickly. The journey only took just over 2 minutes. About a minute into the trip, SAROS spoke the following:

"Your destination is Venus, the second planet from the Sun. Its average distance from the Sun is 67,237,909 miles. The diameter measures 7,520.8 miles. The length of day is 243.69R Earth days. Its orbital period measures 224.67 Earth days. The average temperature is 866.9 ºF: atmospheric gases, carbon dioxide, and nitrogen. The surface gravity measures 90.3 % of Earth's. For other important information, please read the planetary book."

Curious to know what R days meant, I looked in the book under the chapter on Venus. I found that the R stands for retrograde. The whole planet is literally rotating on its axis backward. I also learned that Venus rotates upside down, and that its day is longer than its year. What a weird place! It was only the beginning of Venus' surprises.

The ship was already descending when SAROS ended the information about Venus. When the Lotania landed and shut down its engines, I saw that it had docked with a structure of some kind. I attempted to speak with SAROS, unsure if it was capable of conversation.

"SAROS, what is the structure for?"

To my continued amazement, SAROS responded with the following:

"The structure is built to protect occupants from the extremely lethal conditions of Venus. Although you will be safe within the structure, you will still need to dress appropriately for the outside conditions of Venus."

SAROS' answer didn't instill within me the confidence I would have liked. Nonetheless, I prepared myself with the proper attire and opened the outer door of the ship. Upon entering the protective structure, I saw a machine that looked like an old miniature golf hole, or a skee-ball game. Off to the right of the entrance, I beheld another door with a red button to its right. There were also specialized windows within the safe structure from which I witnessed periodic rain bursts, which caused the ground to steam. I later learned that this was because the liquid falling from the sky was sulfuric acid instead of water. I also saw unpredictable lightning bolts striking the surface. All these conditions, combined with Venus' normal average surface temperature of almost 900 degrees Fahrenheit, made it seem like this planet would be the last I would live to encounter. It wasn't exactly my most cheerful thought, but when the conditions outside are worse than those of hell, it's okay to worry a little.

I forgot to mention that off in the distance, I also saw another protective structure amidst the steaming barren and rocky surface. I was a bit preoccupied with the thought of how Venus was about to destroy me. Would I be seared to a crisp, crushed by the intense atmospheric pressure, eaten away by the sulfuric acid, or all of them at the same time? I mean, seriously! I read that the rocks are flattened due to the incredible surface pressure on Venus. At night, they glow a dull red color because of the extreme heat.

Alright! Enough gloom and doom. I'm going to be fine.

I figured that inside the other building was the paper fragment I was looking for. The question of how to get to the distant structure unharmed remained unanswered. With all this in mind, my attention was focused on the machine resembling a game of some type. I noticed that next to the skee-ball-like machine were written instructions that read:

"Get 1,800 points or more with three balls for safe passage to the structure beyond. If you fail, you're on your own. You only get one chance. Good luck!"

I gazed mindlessly at the instructions while reviewing the message's compassionate words. I then saw a putter resting upright against the machine. Next to the putter were three balls. I looked over the skee-ball-looking object, and saw it was constructed like a mini golf hole. It had a flat, skinny path like that of an old pinball machine. The path then led to a ramp, which entered into a spiral board consisting of eleven loops, each carrying a different point value from zero to one thousand. The different point values were listed at the trough of each spiral loop. My job was to shoot three balls into the spiral board and hope to get 1,800 points or more. My reward would be to live and see another day. No pressure there!

Well, there was no sense in waiting. I picked up the putter and placed a ball on the launching surface. I hit my first ball, aiming for the very center, where a value of 700 points awaited me. The shot went well as the ball rolled to the center of the spiral. When it got there, the speed of the ball, along with the construction of the spiral's center, caused the ball to jump out of the 700-point area and into the 300-point hole beneath it. Things were already beginning to look a bit dismal. The feelings of crushing pressures and searing heat began to harrow up in my mind again.

Okay, two more balls to shoot. I tried again for the center, hoping to control the strength of the ball. This time, I succeeded in getting the ball in the center and received 700 points. I now had 1,000 points with a final ball to shoot. The 1,000-point loop was under the 300-point loop.

The 800 and 900-point loops were spaced lower in the spiral. I needed a strong enough, but well-controlled putt to make it to the 1,000-point loop.

Preparing myself, I struck the ball, but not before accidentally striking the ground first. The ball took off without sufficient strength to make it into the 1,000-point loop. I could only hope it would land in either the 800 or 900-point loop to give me the points necessary to grant me safe passage to the other structure. The ball entered the spiral and looked like it would make the 900-point loop. It climbed the loop heading for the desired points, but did not make it all the way. It rolled back, settling itself in the 600-point loop beneath it. The 800-point loop was beneath the 600-point loop. I felt my heart sink, my hopes disappear, and incredible fears wash over me all at the same time. I didn't make it. I only got 1,600 points. There would be no protective connection between buildings to aid me.

I now had a choice to make. I could continue and face the fatal conditions of Venus, or I could return to the ship and leave Venus for another location. I could always come back later, but it wouldn't matter. The instructions beside the machine said I would only get one shot at success, and I blew it. Returning later would only mean facing Venus under the same existing conditions. Of course, I could always just return to Earth, but I'd never be able to live with myself if I did that. So, I either stay or come back later. I, therefore, decided to stay and deal with Venus, and hope that fortune would favor the unfortunate.

Taking a deep breath, I walked over to the door and pressed the red button off to its side. I then heard a voice that greatly startled me, which said,

"There is a twenty-minute limit for anyone exposed directly to Venus' atmospheric conditions."

Then the door to the protected structure opened, and I stood there with widened eyes, gazing out over a barren landscape that shimmered naturally due to the incredible heat emanating from its surface. About seventy yards of sulfuric acid downpours and unpredictable lightning strikes stood between me and my destination. I had twenty minutes to live after going outside.

I stepped out the door and began to run. I wasn't thinking about leaving a first footprint this time. I found it was quite easy to get around on Venus due to its gravitational pull being very similar to Earth's. My luck being what it is, it began to rain no more than ten seconds after leaving the structure. Immediately, I saw the steam rise off the ground as I did before. This time, *I* was also included. I noticed my suit was also steaming, and the fabric was bubbling up. My suit was being dissolved right in front of me. I continued to run, hoping to find a place to seek refuge from the rain. Ahead of me, off to my right, was a sizable crack in a boulder, which I slipped myself into. I remained there until the rain subsided. Luckily for me, the short-lived downpour ended as fast as it began. When I felt it was safe to leave, I continued my run toward the safe shelter.

I kept an eye on my suit, which continued to sizzle from contact with the sulfuric acid rain. About 25 yards to go, a lightning bolt struck the ground about an eighth of a mile off to my left. The resulting thunderclap was very near deafening, and I lost my balance as I ran. I felt myself sweating inside my suit, which probably had nothing to do with the conditions of Venus. Finally, I reached the door to the other structure and went inside. Right after I did, it began to rain again. I don't know if my suit would have been able to take another drenching.

"Oh, what's the difference?" I thought. *"I may have avoided the second downpour, but I still need to return to the original structure where the Lotania is docked."*

Having made it to the second structure with less than fourteen minutes to go, I noticed that my suit had stopped bubbling. The dissolving had stopped.

The second structure wasn't half the size of the docking structure. It had very little in it, but no matter, it had what I came for. On a table in the center of the room was a metal box. I went over to it and opened it. Inside was another paper fragment similar to the one I found on Mercury. It had the same unknown language written upon it. I safely stored the fragment of paper in my half-eaten backpack, hoping to see if the pieces would connect together when I returned to the Lotania. That is, if I would return to the Lotania.

To be safe, I waited a little while and let the next downpour come and go. When it passed, I took off for the ship with less than seven minutes to go. On my way back, I almost managed to avoid the next rain. I would have succeeded if I hadn't tripped over the rough terrain. Despite my clumsiness, I was only exposed to the rain for about ten seconds before entering the original shelter.

Once inside, I headed back to the Lotania where I immediately removed the suit and boots, hoping to avoid any possible problems to my own clothes. SAROS immediately instructed me to place the clothing into a safe area where they would be properly treated to avoid further contamination.

When the excitement was over, I took a few moments to look at both fragments of paper side by side. There was no doubt that they were part of the same document. Unfortunately, I wasn't able to tell if they linked together. The edges didn't match, so my gut said they didn't, and probability agreed seeing I only had two pieces out of twelve.

It was time to move on. I survived what I hoped would be the worst this adventure would dare to offer. I never want to see Venus again, except through a telescope from my backyard. I sat again in the command chair and heard SAROS ask me to input my next coordinate. The next place I chose to go was Earth's satellite, Luna. I typed in the number 2298446, took a deep breath, and pressed the engage button.

CHAPTER 6
LUNA'S ICE AND SNOW

As I approached Earth's nearest neighbor, I thought to myself that it had been about 25 years since humans last started visiting the Moon again. The Artemis III - V missions took place in the latter half of the 2020s, eventually landing near the region of the Moon's south pole. Before the Artemis missions, man had not been on the Moon since 1972, with the Apollo 17 mission.

As I descended to the Moon's surface, I realized how far off movies really were when they simulated a lunar landing. The realism can't be put into words any more than pictures can do justice to a beautiful panoramic landscape. Freeing me of my thoughts, SAROS spoke up while the flight was nearing its end.

"Your destination is the Moon, Earth's natural satellite. The average distance from the Earth is 238,855 miles. The diameter of Luna is 2,160 miles. Luna's length of day measures 27.32 Earth days. Its full phase to full phase cycle length measures 29.53 Earth days. The average temperature is 240 °F (day lit side) and -290 °F (dark side). Atmospheric gases are essentially a vacuum. Luna's surface gravity amounts to 16.5 % of Earth's. The Lotania will safely land on the Moon's surface near the Sea of Tranquility. For other important information, please read the planetary book."

The whole trip only took a couple of minutes. Before I knew it, I was dressing again to go in search of a paper fragment in another hostile environment. I was able to wear the normal space suit this time, but my boots needed weight due to the low gravity. I was about to weigh one-sixth of my normal weight. This would be the lowest gravity I'd experienced up to now.

Once properly dressed, I opened the ship's outer door and stepped onto the lunar surface. Again, I was awestruck. The stars were brighter than can be imagined. Despite the starkness of my surroundings, there was untold beauty all around me. I was most amazed when I turned and saw the Earth seemingly afloat in an ocean of night. Tears welled up in my eyes as I became emotionally overtaken. I can't say how the astronauts of the past felt, but if their feelings were even remotely similar to mine, I know their trip to the Moon was totally worth it. The Earth was indescribably beautiful.

After taking in the incredible scene, I looked around at my immediate area and saw a couple of machines a little way off on the dry and cratered landscape. As I approached the machines, I came across a plaque with writing which read:

"Why hide thy face O lesser light, thy position disagrees. Shine forth again, dear orb of night; thy beauty will appease."

I couldn't say I had any idea what the words meant, although they were rather pretty. While thinking over the words, I took a closer look at the machines to better understand them. One of the machines stood about four feet high, looking like a clock on top. But instead of having twelve numbers on its face, it had eight circular indentations with a number under each one. The numbers were 0, 45, 90, 135, 180, 225, 270, and 315. The 0 was placed at the bottom, followed by the 45 to its right in the 4:30 position. This placed 180 at the top.

I also saw, about a foot below the clock face, a small compartment. Inside the compartment were eight discs. On each one was a picture of the Moon in one of its phases. Assuming the numbers under the indentations to be the degree positions of the moon phases relative to the Sun and Earth, I placed each disc within its proper indentation. Of course, I used the book to help me know how to do this.

I began by placing the new moon phase in the 0-degree position. According to the book, the 0-degree position occurs when the Moon is between the Earth and the Sun. From there, I worked my way around filling in each indentation, ending with the waning crescent phase in the 315-degree position. When I correctly placed the phase discs, all of the indentations lit up. The indentation in the 180-degree position blinked on and off. I hesitantly and carefully touched the blinking 180-degree indentation, which activated the nearby second machine by lighting up the Sun on what I assumed to be a model representation of the Sun, Earth, and Moon.

The Sun sat on top of a metal pole with a rod protruding from it. On top of the rod was the model Earth. From the Earth, another rod supported the model moon. The Earth's rod was designed to orbit the Sun, while the Moon's rod was designed to orbit the Earth. The question was, what do I do now? I had two machines, one with a blinking light in a specific location. I had a Sun, Earth, and Moon model with the Sun lit up, and a plaque with a poetic verse written on it. *"How do they all tie together?"* I wondered.

I looked through the book for information on the Moon's phases to see what was so significant about position 180 degrees. I learned that a full moon occurs when the Moon is in this position relative to the Earth and the Sun. If the three bodies line up with each other at this time, a lunar eclipse occurs. According to the book, a lunar eclipse can only occur when the Moon is full, and the Moon can only be full when it is 180 degrees away from the Sun. This must be why the 180-degree indentation was blinking on the first machine. All of this was beginning to make my head spin.

The information was all well and good, but it didn't tell me what I needed to do. I assumed that I could move the arms of the model around to simulate a full moon in the 180-degree position, but *"where does the poem fit in?"* I thought.

Looking over the writing again, I wondered if a lunar eclipse was being referred to within the poem. It spoke of the Moon hiding her face, although her position does not agree. Could it mean that normally, when the Moon is 180 degrees from the Sun, a viewer usually sees a full moon? If so, the normal view of the Moon in a 180-degree position is a full moon. The only time an eclipse of the Moon occurs is when the Sun, Earth, and Moon line up precisely on the solar system's orbital plane. Therefore, the poem is correct in saying that the Moon's appearance and position do not agree during an eclipse. The Moon's face should not be darkened when in the 180-degree position. That is why the position does not agree. Well, if nothing else, I was learning a pretty good amount of astronomy.

With nothing to lose, I decided to give my earlier idea a try. I arranged the model to simulate a full moon/lunar eclipse. I placed the model moon in the 180-degree position. After doing so, the lit Sun on the second machine projected a beam of light onto the Earth. The Earth, in turn, then projected an image onto the eclipsed Moon. The image projected was a seven-digit number, a coordinate that read 2298447. Curious to know where the coordinate would lead, I immediately returned to the Lotania. Once aboard, I input the seven digits and pressed engage.

The craft only traveled for about 12 seconds when it came to rest again. The coordinate corresponded with a latitude and longitude of 89.9 degrees south, and 0.01 degrees east, respectively. SAROS then said the following:

"The Lotania has landed near Shackleton crater at the Moon's south pole. The crater measures 13.05 miles in diameter, with a depth of 2.61 miles. While looking at the Earth in space from this location, the Moon's South Pole lies on the rim of Shackleton crater at the 10:30 position. Some of the crater's rim is continuously in sunlight despite the Sun's low altitude above the lunar horizon. These peaks have been called "The peaks of eternal light."

When SAROS finished, I found myself wondering why the Lotania was sent to the Artemis mission's old landing site.

When all the engines shut down, I exited the ship. The temperature had changed drastically from 240 degrees Fahrenheit, to about -298 degrees Fahrenheit. To be on the safe side, I re-entered the ship and changed into my suit that handles extreme temperatures.

After exiting the second time, I noticed it was quite a bit darker in this location. This was due to the Sun barely being visible above the horizon. I pulled the flashlight from my travel bag to illuminate my immediate area. Looking around, I caught sight of the Earth again, only this time, it was upside down. Over the time I spent at the South Pole, I noticed that the Earth also appeared to rotate backward. The amount of rotation was small, but discernible. This was all totally mind-blowing and fantastic.

Breaking myself free from the majestic sight, I wondered what was so special about this particular crater. It was then that my flashlight beam fell upon what appeared to be snow. *It couldn't be*, I thought.

Whatever it was, it was close to the crater's rim. Carefully, I approached the crater and stepped into it just enough to see what the substance could be. I was totally astonished to find out that it truly was ice and snow. *"How did it get here?"* I thought. *"There is no atmosphere to speak of that could cause weather patterns to occur."*

It wasn't until later that I learned that a comet was believed to have impacted the Moon's surface near the South Pole a very long time ago. This conclusion was brought about by radar tests, which indicated the presence of ice within the crater. This must be why the Artemis mission was chosen to come to this location. It was hoped that man could use this area as a pit stop to prepare for locations farther away. The ice would be a source of water. Well, it seemed that I had just re-confirmed, the comet hypothesis.

Truly, this adventure was great; dangerous, but great.

Okay! I just found what could be the remains of a comet within a crater located at the Moon's South Pole. *"I guess that would make this crater pretty special,"* I thought. I continued sweeping the flashlight over the barren landscape to see how large an area the ice and snow covered. As I did so, the flashlight beam reflected on something else to the right of where I approached the crater. It was a metal box about four feet down inside the crater. I had no problem in retrieving the metal box. The initial slope of the crater near its rim was quite shallow.

After exiting the crater, I took one last look around. *"What is all that?"* I thought. My flashlight beam landed on a set of structures off in the distance. They were obstructed by a sizable boulder not far from the crater. Between the darkness and the boulder, it's no wonder that I didn't see what turned out to be the Artemis mission's moon base. Appearing to be abandoned, I left the area undisturbed out of respect.

With the metal box in hand, having temporarily forgotten all about it, I returned to the Lotania to see it better. Once inside the craft, I opened the box to find my third fragment of paper with the strange alien writing. Again, I wasn't certain, but this time I felt that two of the papers could possibly fit together. The edges of the two pieces joined nicely. Seeing progress, I safely placed all the paper fragments within the box just retrieved from the Moon's surface. After putting my suit and footwear back in storage, I returned to the command chair, where I input the next coordinate, 3167208, for the red planet Mars. Part of me was sorry for not visiting the old moon base. Perhaps, one day, I would return.

CHAPTER 7
THE SLEEPING GIANT OF MARS

I was beginning to feel more comfortable with my travels through space. Considering all I've dealt with, I felt as much at ease as I could. As I rapidly approached Mars, SAROS gave its report as it had previously done when approaching Mercury, Venus, and the Moon. The information given was as follows:

"Your destination is Mars, the fourth planet from the Sun. The average distance from the Sun is 141,633,260 miles. The diameter of Mars is 4,220.6 miles. The length of the day measures 24.62 hours. The length of the year measures 686.97 days. Mars' average temperature comes to -81.7 °F. Atmospheric gases contain carbon dioxide, nitrogen, and argon. Mars' surface gravity measures 38% of Earth's. The Lotania will safely land at the base of the solar system's largest mountain/volcano, Olympus Mons. For other important information, please read the planetary book."

Two and a half minutes after takeoff, the Lotania landed safely on Mars. While descending to the Martian surface, I was surprised to see a metal box, which may have contained the paper I sought. It sat atop the famous Martian land feature called Olympus Mons, a shield volcano known to be the tallest structure in the solar system. Olympus Mons measures a height of 16 miles and has a base diameter of 374 miles. Due to the volcano's steep slope, the Lotania landed near Olympus Mons' base. I also saw, as the ship descended, a very small area of flat and level land a little ways beneath the summit. It turned out that the small area could fit the ship. But I'm getting ahead of myself.

After landing on Mars and checking its surface and atmospheric conditions, I proceeded to get dressed into the normal space suit and weighted boots. Having weighted shoes sure helped me walk a lot easier. Once dressed, I exited the ship and stepped onto the red sands of Mars. Even though I recently visited other planetary bodies, I found myself still in a state of wonderment with each new experience. Looking around, I saw a barren surface with boulders strewn about. The rocks were covered with rusty red soil. The sky was clear and bright but orange-tinted in appearance. There were a number of craters to be seen, but nothing in comparison to the surfaces of Mercury and the Moon. The thin atmosphere of Mars saved it from some of the impacts it might have received had it been void of an atmosphere.

This was truly a remarkable moment. Still, I can't say that my next decision was one I care to remember. I had to decide how to ascend the solar system's tallest structure. The idea of climbing 16 miles to a volcano's summit was by no means appealing, but what else could I do? Of course, there was that flat and level land area near the summit. I could attempt to land the craft on that area of land, but that would mean flying the ship manually. I wasn't sure I had enough pluck to try that yet.

While making up my mind about which choice to make, I walked around the landing site. Then, a couple of questions occurred to me.

"*Why this particular spot near the base of the volcano?*" I thought. "*Why not the other side?*"

The ship usually landed in an area that was beneficial to the mission. The Dearlings, or whatever Richard called them, were testing intelligence and other qualities. With all this in mind, I began examining my immediate location in greater detail. Sadly, I couldn't find anything that stuck out strangely, except a peculiar protrusion of rock that actually did stick out of the volcano's base. I walked over to it and saw a boulder blocking what appeared to be an opening in the volcano's base. I tried to squeeze myself through a small gap to get past the boulder, but no luck. It was time for me to cut back on the bagels.

Somehow, I had to remove the boulder. If only my gear had some C-4 or TNT in it. But alas, I had no such items. I began heading back to the Lotania to work out this dilemma, when I only walked a few feet and had the answer. The Lotania's rover! Its robotic arm should be able to move the boulder.

I quickly returned to the ship and opened the back of the craft so the rover could get out. Entering the Land Rover, I drove it out of the ship. I did a little test drive with it before performing my task at hand, seeing that I sort of hit the wall of the Lotania while backing the rover out. No worries! There were no damages to the Lotania or the rover.

When I was more comfortable and confident with the rover, I drove it over to the boulder and gripped it with the rover's robotic arm. The rover was much stronger than I thought it would be. It dislodged the boulder from the cliff as though it only weighed a few pounds. With the boulder removed, I saw that the opening was an ancient volcanic shaft. I also luckily saw that the back of the boulder had something etched into it. The etching was simply four distinct grooves colored black. It obviously meant something, but I had no idea what.

I needed my flashlight to continue into the shaft. About forty feet in, the path split in two. One path broke off slightly left and continued at the same level. The other path veered to the right and ascended. Continuing along the level path turned out to be no good. The path became impassable due to rubble from a cave-in. I returned to where the paths split and began my ascent along the path to the right. I wouldn't have gone wrong if I had remembered that song I learned in church sooner. It is called "Choose the Right." Well, No matter, I'm on the correct path now.

The ascending path leveled out for a small distance after walking about 50 yards. It was here that I found another Land Rover similar to the Lotania's. Looking over the vehicle, I saw that the Land Rover's driving controls and functions were the same. Therefore, I happily got into the rover and continued up the path. The trail steadily climbed for the next ten minutes when I came to another area that leveled off. This time, the area was circular and much larger. What I saw before me were five additional openings to paths, not including the one I just exited from. The paths were spaced like the numbers 8, 10, 12, 2, and 4 on a clock face, with me in the 6 o' clock position.

Having no idea which way to go, I took a chance and tried the path to my immediate left. The path had many twists and turns and appeared to be going nowhere after a few minutes. There was no area to turn the rover around, so I slowly backed out of the passage. This took a good twenty minutes to accomplish, but I finally made it out after bumping and scraping the rover several times.

With one path down and four to go, I tried the opening on my immediate right from where I entered. Along this path, I found myself heading into a dark zone. I stopped the rover, unable to see well enough with just the headlights. In need of more light, I switched on the rover's exterior illumination, and was glad I stopped. Not far before me were two things: a dead end in the form of a wall, and large volcanic-dwelling organisms. I had no wish to overstay my welcome, so I hi-tailed myself out of the cave as fast as the rover could get me out while moving in reverse. Of course, one of the creatures began pursuing me, but stopped when it reached the edge of the dark zone. I'm guessing the creature was sensitive to light, luckily for me. I must have irritated it with my exterior illumination.

Once out of danger, I exited the path back into the circular area.

"*This is ridiculous,*" I thought. "*This shouldn't be this difficult.*"

With three paths left, I sat back and thought the problem through. About a minute later, I slapped myself in the head.

"*That's it!*" I said aloud.

I remembered the boulder. That is what the notches were for. There were four notches in the rock. I'm supposed to take path number four from where I entered the circular area. I just had one more problem: the fourth path moving clockwise from where I entered, or counterclockwise?

Call it intuition or whatever you want, but I'm dealing with an intelligent and ordered race of beings. I therefore chose clockwise, seeing it was the natural order of movement.

Continuing onward, I entered the path in the 2 o'clock position. The rover began to ascend again, which I took as a good sign. I continued my climb for another seven minutes when I saw light ahead of me. It came from an opening in the volcano's fifty-mile-wide caldera.

Not much further along, the path ended suddenly. Continuing forward would lead to the bottom of Olympus Mons' lava pool, *very* far beneath me. Not able to drive any further, I continued on foot. I came to the path's edge and looked carefully into the abyss. When I said it was a long way to the lava pool beneath me, I had no idea just how far. The distance was mind-blowing. I became dizzy while looking and had to back away from the edge. After giving myself a few moments to feel better, I caught sight of a thick rope ladder hanging down off to the right side of the exit. Satisfied with the ladder's condition after looking it over, I began climbing it.

The ladder led the rest of the way up and out of the volcano. After exiting, I saw that twenty feet or so away was the metal box I saw as the ship descended to the planet's surface. Once reached, I opened the box to find the paper fragment with the mysterious writing. I safely stored the paper away with my gear and began returning to the Lotania. Pausing a few moments, I took it upon myself to look around from the solar system's highest peak. The view before me would not allow words to do it justice. From this height, I felt as though I could see halfway around the planet. It was truly a breath-taking experience. After a few minutes, as if waking from a trance, I turned myself away.

As I approached the ladder, I noticed an excess amount of steam rising from the volcano. While descending the ladder, Olympus Mons shook rather violently. Holding on very tightly, I looked down, although foolish, and noticed what seemed to be a bit more activity than when I ascended the ladder.

"*This shouldn't be happening",* I thought. According to the book, Olympus Mons has been labeled an extinct volcano."

Once the quake subsided, I descended the rest of the ladder and returned to the path where the rover was parked. Having enough room to turn the vehicle around this time, I headed back to the circular area of openings. Driving was a bit more difficult on the way back due to aftershocks and rubble on the path from the recent quakes.

At the entrance to the circular area, I found a boulder half-blocking the way in. I removed it using the rover's arm and was soon on my way again. Once inside, I had to remember the correct path to get back. Using my location as the new 6 o'clock position, I needed to go to the 10 o'clock position. Getting to the correct path wasn't easy. Large rocks had to be driven around, whereas others had to be driven over when possible.

Finally making it to the correct path, I drove the rest of the way down the mountain, having to stop a couple of times to remove boulders off the path. Near the end of the road, another major quake began loosening rocks all over the place. One of them hit the roof of the Land Rover, damaging the robotic arm. Thankfully, I did not need it anymore to reach my destination. At the end of the trail, I exited the rover and ran to the opening of the volcanic shaft, back onto the surface of Mars.

Under the Martian sky again, I went back to the Lotania. Off to my left, I saw the Lotania's rover that I completely forgot about. Quickly running over to the rover, I entered it when another massive Marsquake began. This time, it began to rain ash. When I looked up, I noticed that wasn't all that was making its way down. I took off for the Lotania faster than I believed possible. I was glad not to have closed the ship's outer door for the rover. I drove the vehicle into the ship as fast as I could safely do it, and then exited it. As I closed the main door of the craft to pressurize it, I noticed lava escaping from the volcanic shaft I had left just minutes ago. Lava bombs were falling around the ship. Having seen enough, I made my way to the control console. I was inputting my next coordinate when SAROS asked me to input a new coordinate.

"I'm way ahead of you, sister", I said.

I input the next number on the list without checking where it would go. The number was 3490075. I pressed engage and happily lifted off of Mars.

CHAPTER 8

PHOBOS: SIGHTS CAN BE DECEIVING

While lifting off, I had no time to check what my new destination was before SAROS came on and said,

"Your destination is Phobos, Mars' largest satellite. The average distance from Mars is 5,827 miles. The diameter of Phobos is an average of 14.2 miles. Its length of day is 7.6329 hours. Phobos' length of year is synchronous with its rotation. The average temperature range of Phobos is 25 °F (day lit side), -170 °F (night side). Atmospheric gases, none. The surface gravity measures 0.0054 % of Earth's. For other important information, please read the planetary book."

Before SAROS finished speaking, the Lotania was making its descent. I suppose that makes sense seeing the ship only needed to travel 5,827 miles. While descending, I felt a jolt coming from the ship.

"SAROS, what was that?"

"The ship's latching cables have been fired into the surface of Phobos to keep it from floating off the surface due to the extremely low surface gravity."

Moments later, the ship landed softly on Phobos. It was secured by the cables as they retracted slightly to take up the slack. I'm glad that SAROS was on top of the situation. I didn't think that such a procedure would be necessary. SAROS anticipated my ignorance, which saved the ship.

"Thank you SAROS for staying alert, which is more than I can say for myself."

"Of course, Captain. Don't be hard on yourself. You will get used to the procedures sooner than you think. Besides, firing the cables is one of my duties when the ship is on automatic."

After what I believed to be my near-fatal landing lesson, I read up on the conditions of Phobos. I saw that SAROS wasn't kidding about extremely low gravity. For kicks and giggles, I was curious about what I weighed on Phobos. I calculated about 0.77 lb. Talk about a weight loss program!

As for dressing properly, I needed a suit capable of handling extremely low pressure; in this case, none. I also needed my weighted boots to help keep me from floating away. Even with the weighted shoes, I wouldn't be able to safely walk around on the surface. I needed to secure myself with a rope, tying one end to myself, and the other to the ship to keep me from floating off into space.

After I attached myself, I opened the door and walked, or I should say, floated outside. I noticed the metal box was within plain slight, not more than 100 feet from the ship. After what I just went through on Mars, this seemed almost comical. I guess the Dearnig people were feeling a bit sorry for me.

Trying to move around, I wasn't certain if I liked the sensation of near weightlessness as I tried to regain some control with every step I took. I found myself tugging on the rope with every attempt I made to reach the metal box. Every tug pulled me back to the ship, where I held on to the craft's outer rail. Looking around in frustration, I saw the black and starry sky all around me. This helped to calm me down and refocus myself. The Sun occupied part of the sky, so I knew it was daytime, at least on Phobos. It was nighttime when I left Earth, exhausted if I remember correctly. But with all the recent excitement, who could sleep?

I began again, gently and carefully walking over to the metal box. With each step, I yelled at myself for not paying better attention in Physics class when Newton's 3rd law was being explained. In other words, with every step I took, I began to float away. Again and again, I pulled myself back to the ship.

I take back what I said about the Dearlings feeling sorry for me. They're probably watching me right now, getting a major belly laugh out of all this.

Alright, I had to think this through.

"*I need to secure the ship to something else to make a path for myself,*" I thought. "*The new rope will act as a rail for me to hold onto.*"

Looking around, I saw a nearby outcrop of rock. "*Firing my cable gun at it should do the trick,*" I thought.

I took aim, shot at the rock, AND MISSED IT! The cable's momentum pulled me rapidly away from the craft, and I drifted out 50 feet until the rope reached its limit. I stopped abruptly, spinning chaotically for a few moments. If the Dearmies weren't laughing before, they certainly were now. I really need to learn that race's name again.

Managing to get hold of the rope, I pulled myself back to the Lotania for what seemed to be the eighty-eighth time. I'm just glad the line stayed secure and didn't break.

My second attempt at hitting the outcrop of rock was a success. I held on tightly to the ship and gave the line a few tugs to ensure it was secure. When satisfied, I slowly headed over toward the metal box. Although the walk was slow, I did finally make it over.

Looking closely at the container, I saw it was secured to the ground with thin but sturdy metal strips. I should have figured that the box would also need to be kept from floating away.

Of course! Now I needed to go back to the ship to get the small hand laser to break the metal bands. Now, I was certain I was being watched and laughed at by unseen beings from the distant reaches of space. Phobos was beginning to make Mars seem like a picnic.

After another trip to the ship and back, which took about twenty minutes, I carefully began to work on the metal strips. This was not easy, seeing I had to do it one-handed. I could not let go of the rope with my other hand. As it was, every time I turned on the laser, I felt myself begin to drift. I finally just kept it on and adjusted myself accordingly. After a good deal of time, I managed to brake the seals, which were holding the box down. Keeping the box from floating off, I opened it with great difficulty. I then noticed my hand laser was escaping. I barely caught hold of it before losing it forever. Remembering that the box was opened, I carefully took the paper fragment out. It even began to float away. This place is a nightmare!

While storing it with my gear, I caught sight of the box floating upward, now out of my reach. After a few moments of watching its steady ascent, I saluted it and returned to the ship. Holding on to the Lotania after my arduous walk, I released all the tethers and stored them away. I then carefully and happily entered the craft, having no desire to look back. Once inside, I laid down and acclimated myself to normal gravity again. I felt like a block of lead after being in such low gravity.

I removed my spacesuit and boots when I felt better, but continued lying there for a few more moments. Phobos only appeared to be easy. It's no wonder why the box was so close to the landing site.

Moments later, sleep had overtaken me. When I awoke, it seemed to be only a few minutes later, but was actually close to five and a half hours. Attempting to rouse myself to full consciousness, I slowly walked to the command chair. I can only imagine what my breath smelled like. If it was anything like the nasty taste I had in my mouth, then I was thankful not to experience it. Let's just say that SAROS is lucky she's a computer that can't smell,…….or can she?

"SAROS, can you smell things?"

"Not in the way you do, Captain. If you're worried about your breath, don't be. I can only analyze the composition of gases. But after analyzing your exhalations, I'm glad I can't smell as you do."

"WOW!, I thought. That's the first time I ever got ripped apart by a computer."

"Thanks, SAROS! It's nice to know that your sense of humor is functioning so well."

SAROS chuckled at me.

When I finally sat down, SAROS asked me again to input my next coordinate. Stretching first with a yawn and a few more rank exhalations, I typed in the number 4591335, which would take me to Vesta, one of the asteroid belts' four largest asteroids. Pressing the engage button, the Lotania began its ascent. As it lifted off, I heard the surface cables detach and retract back into the ship. Again, I was glad the ship was on automatic. If it was set on manual, I grimace to think about what damages I probably would have caused by not firing the cables when landing the craft.

CHAPTER 9
VESTA: ASTEROID #4 NO MORE

Shortly after the takeoff, SAROS gave Vesta's information.

"Your destination is Vesta, the third largest asteroid in the solar system, and the brightest asteroid as seen from Earth. Vesta's average distance from the Sun is 219,542,096 miles. Its average diameter is 326.5 miles. Vesta's length of day measures 5.342 hours. Its length of year measures 3.63 years. The average temperature of Vesta is -190 °F. Its atmospheric gases, none. Vesta's surface gravity measures 2.55 % of Earth's. For other important information, please read the planetary book."

Still a bit groggy; I didn't catch a lot of what was said. I just remember thinking that approaching Vesta was a bit different. It never got very big like the planets did when they were approached. I didn't even think about the fact that I just left Phobos, which is a very small, potato-shaped object. Did I mention that I was still groggy?

Anyway, the Lotania began making its descent, and I heard the cables fire out of the ship again. *"Oh no! Not another adventure like Phobos,"* I thought. It seemed that I was in for another queasy excursion. Like my previous destination, Vesta also has a very low surface gravity. I will be required to wear weighted boots again.

After getting properly dressed, I went out in search of the fragmented map, which would lead me to an alien world. With a surface gravity of less than 3% of Earth's, I secured myself once again to the Lotania before exiting the ship.

Looking around my immediate area, it looked a lot like Phobos did. This brought me no comfort whatsoever. It was barren, cratered, and dark. I will say that the Sun appeared a bit smaller, though.

I tried walking a little to see what I was up against. Although not as bad as Phobos, I still began to lift off the ground with every step taken. I, therefore, took out the cable gun and looked for an area of rock to fire at. It was then that I saw, hidden in a little area of shadow, a metal box resting on the ground near an entrance to what seemed to be a cave. The entrance was sealed with what looked to be a door that closely resembled an elevator, only made of rock. On the right side of the entrance, I saw a sign embedded within the cave wall. From my distance, I could not tell what was written on it. Below the writing, I saw a lighted rectangle about the length of a person's hand. Figuring that the metal box might also be sealed, I checked to see that I had my handheld laser should I need it.

I fired the cable to the right of the sign, which firmly planted itself into the cave wall. Securing the gun to the ship's outer hull, I made my way over to the cave entrance. It took about 5 minutes to travel the distance. Every time I stepped too hard, I began lifting off the ground. I would pull on the cable to bring myself back to the surface.

Once at the cave's entrance, I went to work opening the metal box. I was right; the box was sealed like the one on Phobos. I opened the box using the laser by removing its thin metal bands. Inside the box, much to my surprise, instead of finding a paper fragment, I found a thin, square metal decoder. The plate was fairly small, with holes punched into it in various places.

			X							
		X								
							X			X
				X				X		
		X							X	
					X			X		
						X				
			X				X			

After looking it over, I placed it carefully with my other gear, and proceeded to examine the writing embedded in the rock face. To my amazement, the message was written in English. It read as follows:

my fRienD:

i am veRy SoRRy to pUt yoU on SUch a

qUeSt, bUt it iS neceSSaRy. yoU wiLL

UnDeRStanD Soon.

Needless to say, the message was very oddly written. The letters R, L, S, U, and D were capitalized, no matter the position of the letter within the word. Like the poem on the Moon, I was sure this also had a meaning that I was yet unaware of. Finding nothing else of interest around me, I touched the lit rectangle, hoping for the best. The door opened rather slowly. I'm sure it was making a good deal of noise that the vacuum of space was not allowing me to hear. When it opened enough, I entered into the cave. Once inside, the door to the cave closed behind me. I must admit, it was a little unnerving having that door close. I felt like I was being entombed.

To my astonishment, as the cave door closed, lights came on. I also found that I was able to walk and stand without floating away. An artificial gravity had been activated somehow. I dared not attempt to remove my helmet, seeing I had no proof of any atmosphere being created.

Looking around, I saw that I was inside a mine. Before me was a mining cart on a railroad track. Next to the cart, on its right side, was a post with a lit green circle on it. I gathered that I was expected to take a ride. Apparently, walking the mine would be too long a journey. I, therefore, entered the cart and looked it over. It had a steering mechanism that resembled a joystick. It seems the track would not be a straight, easy path. Apparently, I would be choosing directions. The only problem was what the directions would be. Up till now, there have always been clues on how to solve the riddles.

I had a decoder with nothing to decode, and an oddly written message. It was obvious that the clue had to be within the message. The message itself was an apology from Mr. Gaber for putting me through all this, as if he was responsible for it all. The only real oddities in the message were the capitalized letters.

"Why those particular letters?" I thought.

I needed directions, but instead, I had capitalized letters.

"Do they relate somehow?" I wondered.

"No"! "It couldn't be that simple," I thought. *"The letters are the directions!"*

"That would mean that R stands for right, L for left, and the S is for straight, I hope. But what directions are U and D?"

"Oh no, you can't be serious! The maze goes up and down as well?"

I needed another look at that message.

I returned to the main entrance of the cave hoping the door would open again. I was very relieved when it did, knowing that I wasn't buried alive. The lights went off, and I began to float upward. I should have been ready for that.

Once secured on the ground after pulling myself back, I took another good look at the message and wrote it down. Securing the message, I re-entered the cave, experiencing everything all over again. Once re-adjusted, I studied the message and realized that the order of the capital letters within the sign were the correct choices I needed to make as the mining cart traveled. This didn't look easy, but I had to try. I carefully extracted the letters from the original message and wrote them down separately. This made it easier to follow.

After taking a deep breath, I touched the green light on the post. The mining cart began to roll forward. I had to stay completely focused; there would be no second cart should I mess up. The intersections came at a very quick pace, about every four seconds or so. It seemed like the longest trip I've ever taken. Finally, the cart began to slow down and eventually stopped. I sat there for a few moments to gather myself. That was the most nerve-racking part of my entire adventure so far. Yes, even more so than Phobos.

After a couple of minutes to calm down, I happily exited the cart. I noticed another post just like the one near the cave entrance. This wasn't a good sign, but it was not totally unexpected. Looking off to my left was a structure that looked like the mining office. I couldn't imagine having to traverse that maze every time I came to work.

Entering the mining office, I could see that there wasn't much to look at. There were a couple of work desks, one on each side of the office. Each desk had a single draw in it, and there were chairs that belonged to the desks. A nicer but well-used cushioned lounge chair rested in the back right corner of the room. A fake plant sat on a small table in the back left corner of the room. There were a few pictures hanging on the walls, along with a clock and two lights that hung from the mining office ceiling. On the back wall, in between the fake plant and the lounge chair, I saw another door with something round on it. The overall appearance of the room was rather dusty, apparently unused for some time. As I approached the opposite door, I saw a combination lock, much like one found on a safe. Above the combination wheel, a written question read:

CAN YOU MULTIPLY?

It seemed I would need a combination. Of course, I could just simply use something from my gear to break in, but where would be the fun in that? That would be my last resort should all attempts fail.

I began looking around the main office for any clues. I tried looking inside the desks, but they were locked. I searched for the keys in various places with no luck. I checked under clutter, which was on top of each desk. I checked under the desks and under the seat cushion to the lounge chair. Nothing! Standing back from the chair, my peripheral vision caught sight of something shiny inside the plant. On the fake soil was the key that unlocked both desks. See, I knew I wouldn't have to break in.

The desk on the left side of the room had absolutely nothing of interest within it. As I approached the second desk, I felt a small quaking that lasted for a couple of seconds. That was unexpected and quite unsettling. After a moment, I continued to unlock the second desk. I found within the draw two important sheets of paper. One of the sheets had the following written on it:

A	B	C	D	E	F	G	H	I	J	K	L	M
1	2	3	4	5	6	7	8	9	10	11	12	13

N	O	P	Q	R	S	T	U	V	W	X	Y	Z
14	15	16	17	18	19	20	21	22	23	24	25	26

3 + 18 = U 20 - 6 = N 4 x 6 = X 16 / 8 = B

G + K = 18 V - Q = 5 E^2 = 25 U / C = 7

This paper seemed rather simple to understand. Each letter of the alphabet was given its number equivalent within the alphabet. The example 3 + 18 = U simply means that 21 = U, which is correct. The letter U is the 21st letter of the alphabet.

Looking at the second paper, I saw the following:

2	T	E	6	N	S	7	A	Q	9
P	N	3	9	B	D	H	4	8	C
6	J	M	2	U	7	T	K	L	E
G	N	8	R	5	1	C	R	X	V
X	2	Q	F	E	3	3	M	5	Z
8	A	V	T	L	6	2	6	E	4
5	B	3	H	S	R	4	I	P	1
W	J	1	Q	D	7	2	Z	M	1

I found myself staring at this paper for a while with no success. What woke me from my concentration was another tremor, only this one was stronger and longer. The asteroid seemed structurally sound, but I wished to stay no longer than necessary. I looked back at the grid of numbers and letters again, but could see no discernible pattern at all. I then remembered again that I'd always been given what I needed to unravel the mysteries of my adventure.

"When will I wake up?" I thought.

I quickly went to my gear and pulled out the thin metal decoder I took from the metal box before entering the cave. Correctly orienting the metal decoder on the paper revealed letters and numbers in the following order:

ENTERR25L6R12.

In other words, ENTER R25, L6, and R12. I had found the combination to unlock the door at the far end of the mining office. After walking back to the door, my attention was briefly drawn again to the question, "CAN YOU MULTIPLY?" before I entered the combination to unlock the door. After inputting the numbers, I imagined hearing the joyful clicking sound of something unlocking. Hoping that the lock did click, I tried the door, and it opened.

Inside the next room, I saw nothing but a small table with a metal box sitting on it. Seeing the metal box brought a smile to my face. Overall, the room itself was not much bigger than the table. On each side of the table, there was just enough room for me to get around it. Lucky for me, I'm not claustrophobic!

Next to the box lay a sheet of paper. Looking at the box, I saw that it was just like all the others I've found up till now, save for the box's top. This box had three vertical wheels, each numbered from zero to nine. Starting to finally get the hang of things, I looked at the paper on the desk, hoping to find the combination. Just then, a third quake began. This time, the walls and ceiling began to let go of loose material. I had to hurry and get this box open.

The paper on the table showed the following information:

C	T	I
5	4	0

L	K	F
7	9	2

N	B	Q
?	?	?

This puzzle wasn't going to stump me. I remembered the question on the door, and the information from the desk paper about letters equaling numbers. I, therefore, multiplied C times T times I, which was 3 x 20 x 9. I got an answer of 540. Nice! The same number was under the C, T, and I in the puzzle. L times K times F equaled 792. Yes! I had it! Multiplying N, B, and Q together gave me a total of 476. Entering the numbers 4, 7, and 6, respectively, on each vertical dial unlocked the box.

There it was! I was another step closer to receiving the information on how to get to the alien world. While putting the paper fragment with my gear, quake number 4 shook my surroundings, sending me to the ground. This was like a magnitude six quake on Earth. This was getting very serious now. Each quake was getting stronger and lasting longer. Larger rocks were being dislodged from the walls and ceiling now. It was time to leave!

When the quake stopped, I got to my feet and hurried back to the mining cart. I wasn't necessarily looking forward to making the return trip, seeing all my directions would need to be done in reverse order. Needless to say, the quaking renewed my motivation. To make sure I'd be ready, I wrote the reversed directions down before beginning.

Once written, I touched the green light on the post, and off I went. As I traveled, a monster quake occurred. Debris was falling all around me. Thankfully, nothing hit the tracks to block my passage back to the Lotania. Part of the ceiling behind me caved in. It is very possible that the mining office was just buried under a great deal of rubble. With all that was going on, focusing wasn't an easy job, but I successfully made it back.

I was quite a bit dustier than when I first started, but considering what could have happened, I had no room for complaints. Exiting the cart, I ran over to the cave door and touched the lit panel located off to its side. The cave door opened, and the artificial gravity began shutting down. Again, I forgot that was going to happen.

I quickly made it over to the cable gun cord before all artificial gravity had been lost. I had to get used to the loss of gravity again as I pulled myself back to the ship. Once back, I detached the cable gun from the ship while holding on with my other hand. I retracted the cable and entered the Lotania.

As the ship's door closed, what ended up being the asteroid's final quake began. As I stumbled back to the control console to lift off, the Lotania could barely stay planted to the surface, even with the surface cables. In fact, the ground separated near one of the cables, setting it free. I quickly input the next coordinate of 5501977 as SAROS was bellowing out danger warnings to me. I hit the cable release and engage buttons, and the Lotania lifted shakily but safely off the surface of Vesta. All of the ship's cables retracted back into position.

As I ascended, I could see Vesta visibly quaking beneath me. About thirty seconds later, I witnessed the entire asteroid break apart into a massive pile of rubble. It was terrifying, yet awesome to behold. I'm fortunate to have made it off the surface with the fragment when I did. I couldn't help wondering though, if perhaps I did something that caused the quakes to happen. I can't see how, but when you think about it, what were the odds that Vesta would begin quaking after all that time? Well, no matter; another death-defying feat accomplished.

I then said:

“SAROS, please place in the Lotania’s record that the asteroid called Vesta broke apart due to seismic instability at;…….hold on, let me check the time. Man, is it really early morning already? At 5:52 am. Put its time of destruction down as 5:52 a.m.”

SAROS then responded: “Noted, Captain.”

You would think that I’d be used to narrow escapes by now, but I guess not. What’s important is that I’m still in one piece, which is more than I can say for Vesta.

CHAPTER 10
THE DEPTHS OF EUROPA

The next place on the list was Jupiter's satellite called Europa. As expected, SAROS came on and spoke the following:

"Your destination is Europa, Jupiter's smallest Galilean satellite. The average distance from Jupiter is 416,878 miles. The diameter is 1,939.7 miles. Europa's length of day measures 3.5512 Earth days. Its length of year is also 3.5512 Earth days, and therefore synchronous. The average temperature measures -276.1 ºF. The atmospheric gases are essentially a vacuum, and the surface gravity measures 13.4 % of Earth's. For other important information, please read the planetary book."

As the ship approached the surface of Europa, I noticed that it became quite a bit brighter in the craft. I looked out the port window and saw an icy world beneath me. Although the sun was quite some distance away, it illuminated the surface brilliantly. When the Lotania came to rest, I put on my normal space suit and cleated-weighted boots. I figured the cleats would come in handy on an icy surface. The surface gravity was slightly less than the Moon's, so I didn't plan on having problems floating away on the surface.

Taking a deep breath, I opened the Lotania's door and walked onto Europa's surface. It was quite a difference beholding a wintry scene after all the barren rocky surfaces I've been on. While walking about, I came across a worn-down sign lying half-buried on the icy surface. To my amazement, I also saw a crossbow leaning against a small ice-covered hill off to the left of the sign. Next to the crossbow were five harpoon darts in a quiver. Finding these objects didn't make me feel very comfortable.

I began looking at my surroundings in greater detail. I could see nothing alarming or dangerous. Focusing my attention on the items again, I saw they all needed to be freed from the ice. Not having the proper tool to do the job, I returned to the ship to get the ice pick out of the ship's tool box. It was obvious those Derwhig people knew what I'd be up against. When I returned with the ice pick, I slowly began working to free the items from their icy prison. When I freed the sign, I saw the following message written on it:

Europa's sum equals seventy-six. Beware of ....,

The remainder of the message was too worn off to be read. Great! That's all I need: another cryptic message and a warning that got worn away. Continuing to ponder on the message, I looked around for more objects on the surface. Not far off, I noticed what appeared to be a sizable crevasse in Europa's icy surface. Carefully making my way over to it, I looked over its edge and saw liquid water inside it.

"The surface of this moon has a temperature of -276.1 degrees °F," I thought. *"A great deal of thermal heat must be generated from within the moon to have liquid water."*

Whatever the cause, I was actually looking upon water under Europa's icy surface. The water began about 5 feet down the side of the crevasse. I had a bad feeling that I would need to enter the crevasse to continue my search. I could see no other items anywhere in my area. Also, I was given a suit that could go underwater. It kind of gave my next move away.

I, therefore, went back to the ship so that I could change suits. Once in my new attire, I went back to the crevasse. I was ready to jump into the water when I realized the probable fatal mistake I was about to make. I had no idea what to expect down there. Is there life? Is it incredibly deep? Are there dangerous currents that will sweep me away? With all this in mind, I returned to the ship a third time and tied my longest rope to the hull and the other end to me. I made sure I had the items I found on the surface with my gear. After all, they were left on the surface for some reason. What that reason is......I felt I was about to find out. The crossbow and harpoon darts must be for my underwater excursion. This thought was even more unsettling than the first.

Feeling better prepared, I took the plunge into the abyss. I sank deeper and deeper for nearly a minute, keeping my eyes open for any problems as I descended. When I made touchdown, I estimated myself to be about 350 to 400 feet below Europa's icy surface. I cautiously looked around for any unwanted surprises. Thankfully, I encountered none. Somehow, our distant sun could still give a sufficient light source as it shone through Europa's icy surface. Still seeing no immediate danger, I scoped out the area at the base of the crevasse wall. Not far from where I landed, I saw what looked to be an open entrance in the wall of the cliff. It was a large square opening that looked man-made. It was too well-shaped to have occurred naturally.

Cautiously, I entered the voided area in the crevasse wall. The area was about fifteen cubic feet in size. Using my flashlight on the right wall of the cave, I saw twenty-six tiles hanging next to each other. The image was distorted due to being underwater. I could see that each tile had a Dignish number ranging from one to twenty-six. On the wall opposite the entrance, I saw a small compartment that was closed, and I could see no apparent way to open it. On the wall to the left, I saw six indentations the same size as the tiles hanging on the right wall. Finally, on the same wall as the entrance, I saw two levers: one to the left of the opening and the other on the right.

On the ceiling, I saw what appeared to be two fluorescent lights, although I couldn't believe my eyes that I was looking at lights.

It was obvious that the tiles fit into the indentations. I sluggishly moved over to the tiles and took one of them off the wall. It seemed to be made of stone, or perhaps a shell. They were flat on one side and domed on the other. Other than that, there was nothing odd about them, except that they were on a satellite, in an underwater man-made cave, some 390 million miles from Earth. Putting all that aside, though, they appeared quite ordinary.

Moving around was very difficult this far beneath the surface. I was trying to reach the indentations when I noticed the levers again. I was reluctant to fool around with them, fearing I would get trapped inside. Nonetheless, remembering that the Dearnies are a friendly race, I tried not to fear. I went over to the lever on my right while facing the door. I pulled on it only to find that it would not budge. It seemed rusted or locked. I, therefore, went over to the other lever and pulled down on it. This lever moved, turning on the lights. I still couldn't wrap my head around there being lights in this underwater cavity. Oh yes, pulling the lever also closed the door.

"How the junk are those lights getting their power?" I thought.

The lights seemed to amaze me more than the fact that I was slowly being entombed. With the door now closing, I had only one thing left I could do other than panic and scream. I returned to the lever I initially tried and pulled on it again. This time, it worked. Pulling the lever down caused the water to somehow drain from the cave.

With the water gone, my movements were a lot easier. In fact, due to the moon's low gravity, my movements were almost effortless. Adjusting to the new conditions, I easily walked over to the indentations. As mentioned earlier, there were six of them. Each was concave and about the size of the tiles. After looking at the tiles and the indentations, I was sure that the tiles were supposed to fit inside the indentations. *"But which six tiles are the correct ones?"* I thought. Falling back on the pattern of how all the other mysteries were solved on this adventure, I looked over all the information I gathered so far. The only thing that made sense was the sign that read:

Europa's sum equals seventy-six. Beware of...

Pondering the words for a while, I finally realized that Europa is a word with six letters. Remembering the puzzle on Phobos, I tried interchanging the letters with numbers. I retrieved the six tiles with E, U, R, O, P, and A. The numerical value of each letter equaled 5, 21, 18, 15, 16, and 1, respectively. Adding up the letters gave me seventy-six, as I hoped and expected they would. I placed the tiles that spelled Europa in their proper order within the indentations on the wall. When done, the small compartment on the wall opposite the main entrance opened up. Inside the compartment, I saw the metal box, which contained the paper I was looking for.

Putting the paper fragment with my gear, I prepared to return to the ship, but first, I had to get out of the cave.

I walked over to the lever that opened the door and pushed it up. Like before, it did not budge. Then, a wave of terror swept over me. I was about to destroy myself. I'm glad that the Dearyings were anticipating my blunder. If that door had opened, an incredible force of rushing water would have entered the cave, slamming me against the wall. I probably would have been killed.

Shakily, I slowly approached the other lever and pushed it up. The cave began to fill with water at a safe pace. It seemed my alien friends wanted me to succeed. I took that as a thing of comfort. When the cave flooded, I tried the right lever arm again. This shut off the mysterious lights that were still blowing my mind and opened the main entrance door, allowing me to safely exit the cave.

Still pulling myself together after my near-fatal mistake, I began my return to where I jumped into the water. I went to grab the rope to lead me back when I noticed that the rope had been severed. This most likely happened when the door to the cave closed. I carefully began the search for the other piece of the rope, which didn't take long to find.

As I went to pick it up off the ocean floor, I caught a glimpse of why there was a crossbow and harpoon darts on Europa's surface. In front of me, off to my right a bit, I saw a sea creature of alarming size. Even at its distance from me, which I guessed to be more than 50 yards, it still looked quite large. It wasn't moving at the time I saw it, and I thought perhaps it was because it didn't see me.

Right! Fat chance of that!

Just then, the creature began moving toward me. I grabbed the crossbow and quickly loaded a harpoon dart onto it. Still approaching slowly, I aimed and fired using the weapon's crosshair. The dart left the crossbow with amazing speed for something fired underwater. A couple of seconds later, the dart rammed into the creature's underside, bouncing off of it.

"This can't be happening!" I thought.

Regardless of the direct hit, the creature continued its approach after a brief hesitation. The creature was of such mass that the dart seemed ineffective. I loaded another dart, hoping for better results, and fired. Again, the dart bounced off the creature's underside. It could have been the hopes of a desperate and terrified man, but it looked to me that the beastie was slowing down, but still moving closer. I felt bad for having to do it, but I loaded what I hoped would be the last dart I'd need. Aiming again, I fired the dart. Like the others, it bounced off the creature's underside. Thankfully, the creature stopped its forward movement and began sinking to the ocean floor, where it came to rest.

Having compassion for the creature, seeing that I didn't come here to harm anything, I carefully approached its massive body. I saw that the brute was only tranquilized. The darts were not meant to completely penetrate the creature's body. Instead, they only pricked it, putting it to sleep. I was thankful I did not kill it. Hoping the tranquilizer would keep the underwater behemoth asleep for awhile, I quickly began my way back to the Lotania. I wasn't about to take any chances that my friend would wake up early.

Continuing where I left off, I found the rope and picked it up off the ocean floor. It led me back to the area where I jumped into the water. I began an ascent up the crevasse's wall and reached the surface about three minutes later. Once on Europa's frozen surface, I continued my way back to the ship. I found that I was having a difficult time moving, when it dawned on me that my suit was freezing after leaving the water. I figured I only had a few moments if I didn't keep moving. Struggling, I made it back to the ship, breathing very hard. I closed the door behind me without untying the rope from the outer hull. There would have been no time before I froze solid. Safely aboard, I removed my suit and boots. I then untied the remaining piece of rope from the suit and stored them away. When done, I sat on the ship's floor, trying to stabilize my breathing.

As I began to breathe regularly, I was curious to see my progress concerning the paper fragments. I put all the pieces I had together to see if they would begin to fit. Having seven of the twelve fragments, the probability was in my favor that they should. I was not disappointed in my desire. I saw that four of the seven showed a relation to the others.

I now had more than half of the paper fragments in my possession, and I allowed myself to feel a sense of accomplishment in this.

Having safely stored the pieces of paper, I headed over to the command chair. Taking my seat, SAROS asked me to input my next destination coordinate. Looking for the next coordinate, my eyes fell upon the warning message again:

Do not input any other seven-digit coordinate than those listed above; the destination is uncertain.

Feeling a bit daring and desirous for a break from the "norm," I randomly picked the number 5829038, input it into the computer, and pressed engage.

CHAPTER 11

THE DOUBLY DISOBEDIENT PILOT: CALLISTO AND CERES

As though I pressed a button that would begin World War 3, SAROS came on speaking the following:

"WARNING! The input coordinate is unrecognized. The destination is uncertain and can only be estimated as the course proceeds. Do you wish to continue?"

I hesitantly responded affirmatively to SAROS and waited to see the destination. I hoped it wouldn't be a one-way trip to the Sun. SAROS gave me probabilities of possible landing sites as we traveled, and then went on to say,

"Currently, there is a 96.7% probability of landing on Jupiter's satellite called Callisto."

A few moments later, she continued:

"Now, it is certain that Callisto will be the destination."

"Thank you, SAROS," I responded, as I let out a deep exhale.

The craft began its descent to the alien satellite below. No information was given about Callisto to aid me in how to prepare for an outside excursion. Checking the book, I read that Callisto's average temperature is about -218 °F, and its surface gravity is about 12.6% that of Earth. This was close to Europa's surface gravity and not too different from our Moon's.

Looking out the window, I saw that the ground appeared shiny over most of its area. Assuming this to be ice reflecting the distant sunlight, I dressed in my normal suit with cleated/weighted boots. My plan was not to stay long. I knew there would be no prize to gather on this rather heavily cratered landscape.

Exiting the Lotania, I began walking around, taking in the sights. Callisto was extremely cratered compared to other surfaces I've recently been to. It was very necessary to watch my footing. I had no desire to twist my ankle. The surface was very interesting, showing craters filled with ice. The place was so battered I even saw craters within craters. I saw outcrops of rock standing up like frozen sentinels stationed as if they were guards. Even the soil looked different; it was darker than the Moon's or Mercury's.

I walked nearly half a mile when my peripheral vision saw something strange off to my right. Focusing my attention in the general direction mentioned, I saw nothing further. I then walked in the direction to see if my eyes were merely playing tricks on me. After all, I was very tired and could have imagined seeing something. After a few minutes of looking around, I concluded that I was just seeing things. I could see nothing strange in any direction I looked.

Just then, another sight was seen off to my left. This time, it was not so much my peripheral vision that caught it. Moments later, I knew what I saw. It was time to leave. The surface of Callisto was experiencing a meteor shower.

I had been seeing nearby impact dust plumes as the meteorites hit the surface. They began to strike the surface at a greater rate. This was quite alarming due to being unable to see the incoming objects. Normally, you would see a streak of light stretch across the sky. Here, no such thing was possible. Callisto has no atmosphere to streak though. The objects just hit the ground, not diminishing in velocity or size as normal space debris would do as it approached.

I hurried back to the ship the best I could as meteorites crashed around me. I was uncomfortable thinking of myself as a running bull's eye waiting to be picked off. As I got closer to the ship, I saw it get hit by a meteorite. I could only hope that there would be no damage from the impact. Arriving back at the Lotania, I entered the craft and asked SAROS for a damage report while removing my suit. SAROS answered:

"The landing gear has been affected. It would be prudent to leave as soon as possible."

Giving no argument, but not wanting to feel ripped off due to being chased off, I input another unofficial coordinate. I typed the number 4871264, and pressed engage. Again, almost immediately, SAROS said:

"WARNING! The input coordinate is unrecognized. The destination is uncertain and can only be estimated as the course proceeds. Do you wish to continue?"

Once again, I answered yes.

As the ship lifted off Callisto, I asked SAROS for probability readouts like I received while traveling to Callisto. SAROS replied:

"Destination is uncertain. We are heading back toward the asteroid belt, where there are literally tens of thousands of possible landing sites. A vast majority of them can be eliminated due to our current trajectory, cutting our landing sites to less than 20,000. The only certainty is that the Lotania is not on any present course to land on any major planets."

I interrupted SAROS and said,

"We're not heading for the sun, are we?"

SAROS replied,

“Due to the size of the sun and our present distance from it, giving a probability would not serve as of yet. I can calculate a 62.8% probability that the dwarf planet Ceres will be the Lotania’s destination.”

About 10 seconds later, which felt much longer, SAROS said:

“There is now an 88.1% probability that Ceres will be the Lotania’s destination.”

“And the sun?” I asked.

“There is less than a 20% chance of the Sun being the destination, and it continues to decrease,” SAROS replied.

Feeling quite relieved, I began breathing a bit easier again. SAROS then continued,

“There is now a 93.4% probability that Ceres will be the Lotania’s destination; 95.2%, 98.7%, it is now certain that Ceres is the Lotania’s destination.”

I was very thankful as the Lotania descended to the surface of Ceres and that I wouldn’t be a roasted pilot. I knew better than to continue pressing my luck with the unofficial coordinates. I was taking unnecessary risks that stressed me out too much.

Again, I did not receive information on my landing site, as though I was being punished for being a bad boy. Suddenly, the craft made a rather questionable landing. Then, it made another and another. The ship was mildly bouncing around on the Cerian surface. The surface gravity was too low.

“That’s right!” “The meteorite impact damaged the landing gear,” I thought. I immediately placed the craft on manual control, stabilizing the ship as best I could. I fired the surface cables, praying they would latch to the surface. Hoping the cables did their job, I pressed the retract cable button to stabilize the craft to the surface. A few moments later, the bouncing stopped, and the ship appeared stable. I asked SAROS for another damage report. For a moment, I thought she wouldn’t give me one as if to say; what should you care, you disobedient lunatic? But I did get one. SAROS’ hesitation was just due to checking over the craft. I guess I was feeling guilty for being a disobedient lunatic. SAROS went on to report:

“There is no further damage to the craft.”

“Can the damaged landing system be repaired?” I asked.

“I will make the necessary repairs right away,” SAROS replied.

Leaving SAROS to the repairs, I read about Ceres in the planetary book. I found out that the surface gravity of Ceres was just less than 3% of Earth’s. As for the outside temperature, I had no idea. IT WAS VERY COLD! As I proceeded to get my suit on again, SAROS spoke the following:

"I am detecting an unknown source of energy radiating from the surface. A high concentration of radiation is also being detected. Close proximity exposure limit: 8 minutes before bodily contamination."

Curious to know what could be emitting the energy source, I exited the ship and looked around. Aside from the surface being very barren, I couldn't see anything odd. I moved away from the ship to get a better view of my surroundings. It was then that I saw a glow at a bearing of 118° relative to the asteroid. I didn't initially see the glow because I exited the ship opposite the glows direction.

Moving around almost cost me my life again. I forgot about the very low gravity and almost floated off as I tried to walk. *"How could I forget after what just happened to the ship?"* I thought.

While beginning to float away, I carefully pulled out the cable gun from my gear and fired at the rail on the ship's hull. This made me float away even faster. Thankfully, I latched on to the rail and reeled myself back in, regaining my stability. If I had missed, it would have been lights out for me.

I tried to move about on the surface by latching on to boulders using the cable gun. After a good amount of time firing the cable gun from one boulder to another, I saw an object giving off an intense light in the distance along the horizon. I found it difficult to look at, even from my distance. I continued going over to the energy source when I began to feel queasy and warm. From my new position, I could not look directly at the object. Not daring to get any closer to whatever it was, I took readings using instruments I brought from the Lotania. I also took visuals of the object using different electromagnetic spectrum wavelengths. Hopefully, SAROS could analyze the information and identify the mysterious object.

It was time to return to the ship. I felt drained of my physical strength, as though the object was taking it from me. For all I knew, it very well could have been doing so. When I returned to the ship, I was completely exhausted. I entered the ship and removed my suit. SAROS then said:

"High levels of radiation have contaminated your suit. You are also measuring higher-than-average readings. The decontamination process will begin immediately."

A beam of light enveloped me and the suit from somewhere. I remained bathed in the light for over a minute. When the beam terminated, SAROS said,

"The Lotania must leave this area within 4 minutes to avoid radiation contamination. It will begin to affect critical operating systems very shortly."

"How about the landing gear?" I asked.

"The landing gear is now operational," SAROS replied.

Not waiting for further information, I sat at the command console and released the cables from Ceres' surface. I put the ship back on automatic control which retracted the cables into the ship. I input the next *official* coordinate on the list. The new number was 6136090, which would take me to Saturn's largest satellite, Titan. I had enough of playing Mr. dare devil pilot. That last adventure almost put me away by radiating me to death. It was time to focus on the mission again.

Chapter 12
Titan's Fuel

The Lotania traveled for nearly five minutes to get to the new location. This was due to the backtracking of revisiting the asteroid belt. The prolonged silence of SAROS made me feel once again like I was being scolded for being foolishly adventurous. I knew, of course, that the silence was due to the extended travel time. About a minute out, it was strangely comforting to hear SAROS speak the following:

"Your destination is Titan, Saturn's largest satellite. The average distance from Saturn is 759,235 miles. Its diameter is 3,199.7 miles, making Titan the second-largest satellite in the solar system. It is second only to the solar system's largest satellite named Ganymede orbiting Jupiter. Titan's length of day is 15.945 Earth days. Its length of year is also 15.945 Earth days. The average surface temperature is -291 °F. Titan's major atmospheric gases are nitrogen and methane. The surface gravity measures 13.8 % of Earth's. For other important information, please read the planetary book."

As the craft approached the surface of Titan, I was given the treat of passing through a thick layer of atmosphere. This was the first substantial atmosphere around a satellite I had encountered up to now. I learned that it was not a common characteristic for satellites in the solar system.

When the Lotania touched Titan's surface, I dressed in the proper attire and exited the craft. To my astonishment, I beheld a lake with some kind of fluid, which I did not immediately identify. I just knew that it wasn't water. In the distance, I could make out four land masses, which I believed to be islands. I had a bad feeling that what I came here for resided on one of them. After all, that is how my luck has been up till now. Even farther away was Doom Mons, a mountain that shared its name with another from an old movie containing a ring that many creatures were after. I think it was something like that; it's been about fifty years since that came out.

The surface of Titan was covered with icy bedrock, having muddy areas, which I analyzed to contain methane. I then knew that it was quite probable that the lake was a body of liquid methane. According to the information I gathered, the lake was barely liquid, seeing that methane freezes at -296.32 ºF.

Looking upward, I saw that Titan's sky appeared as an orange-tinted haze. Due to Titan's atmospheric pressure, methane rainfall was very probable. From the amount of liquid methane in my immediate area, I figured it rained quite often. I hoped that I would not get caught in the next downpour.

As I investigated the surface, the first thing to catch my eye was the amazing sight of a boat tied to a dock. Being fully acclimated to my adventure, I looked at the boat as though it was perfectly natural. The boat had nothing with it, not even a paddle. From the look of it, I was certain that it needed to be propelled by a motor.

Not far from the lake was a sign, which read:

"Parents of satellites - most to least."

This was no doubt another puzzle for me to eventually solve.

While walking around the landing site, a little way beyond the sign, I found a can of gas, the motor to the boat, and a small metal box containing a key. I assumed it to be the key to the boat, which I soon saw was a correct assumption. I hoped the metal box would contain my paper fragment, but that would have been too easy. The last item I found was atop an area of land that rose above the beach that overlooked the methane lake. There, lying on the ground, was a plaque displaying a line diagram making a pattern of some kind. I tried to draw the pattern on a pad from my gear, but the suit made sketching anything very difficult. Having everything needed to take a trip, I head for the boat to investigate the islands. The question was which island to go to first. I did not wish to play the game of Eeny, meeny, miny, moe, seeing that I'd probably pick an island that would flame me out; so I ended up deciding to go to the first island on the far left.

I applied the motor to the back end of the boat and filled the boat's motor with gas. Untying the boat from the dock, I started the engine and steered toward the mysterious island. Methane or not, the boat traveled smoothly as if it were on water. The engine had to be quite special, seeing it was able to run in conditions so different from Earth's. About a minute and a half later, I reached the shore of the far left island. I knew exactly where I was expected to go, seeing there was also a dock on this island. I took this to be a good sign. After docking, I tied the boat to the dock and exited the craft.

This island's eye-catching sight was a high peak that stood near the island's center. My prize was probably on the top of the hill, and I would, without a doubt, face several death traps on my way up. While having this cheery thought, I noticed a path leading in the general direction of the tall hill. This only verified my guess of where I would find my fragment of paper. The path was fairly straight forward, with very few obstacles to overcome. It wasn't until I reached an area where the path ended due to a crevasse that I had any trouble. The path could be seen to continue on the other side. Unfortunately, the distance was too great to simply walk over. It had to be about 15 to 20 feet from one side to the other.

There were no objects to fire my cable gun at, so swinging over it was not an option. I did not wish to simply fire the cable at the wall of the icy crevasse. There was no guarantee that the cable would penetrate the crevasse wall, let alone hold my weight, while I swung to the other side. I was at a loss for how to proceed. As I pondered the problem, I realized something I had somehow taken for granted. It came to me as I walked about the island. I watched the ground, noticing my steps and how I took each one. I was moving about as though jumping and gliding with every step. *Of course! With enough speed and a hefty jump, I should easily be able to clear the crevasse,* I thought.

Ready to test my theory, I stood a good ways back from the edge of the crevasse. After gathering my nerve, I took off running as best I could. I was not impressed with my speed and felt my jumping attempt would end with unfortunate results. Nonetheless, when I came to the edge of the crevasse, I jumped hard and closed my eyes. I had to open them when I realized I wasn't making contact with anything solid. I figured I was falling into the deep abyss, soon to become a permanent addition to Titan. Instead, to my great relief, I was still soaring through Titan's atmosphere, having jumped over the crevasse, which was now a good 50 feet or more behind me. It was a very neat sensation traveling in this manner. I then realized that I was descending again and that my velocity would not allow me to land very nicely. I was about to make contact with the surface and roll across it for quite a distance. Hopefully, my suit would not get damaged.

The landing ended up not being as bad as I had anticipated. I rolled around a bit but gained control easier than I thought I would. Overall, it was not a bad experience, and my suit was not damaged. Now safe on the other side of the crevasse, I continued toward the high peak at the island's center. When I arrived some 10 to 15 minutes later, I saw that climbing would not be too hard. There were plenty of places to get good footing and hand holds. As I began my climb, it started to rain. I didn't think much of it until I got that horrible feeling of having difficulty moving, just like I had on Europa. The near-freezing conditions for liquid methane were taking its toll on my suit.

Looking around, I found an overhang a little further along that would have to serve as a shelter from the rain. I stood there for some time, waiting for the rain to stop. As I waited, I brushed away areas of frozen methane crystals from my suit.

When I could safely continue, I reached the top of the hill. I saw a pretty nice view of the island below me. I also caught sight of the nearest island to me. What I didn't see was a metal box containing a paper fragment. Needless to say, I wasn't my happiest at the moment. After a few more moments of complaining and sulking, I climbed down the hill. I looked around its base and saw no further sign of a trail. I guess I picked the wrong island and wasted my time. While walking back, I thought about how close I came to being turned into a methane popsicle. None of my thoughts uplifted me in the slightest.

I returned to the boat, but not before seeking shelter from another downpour. Untying the boat from the dock after the storm, I returned to where the Lotania landed. I don't know what I would have done if it had rained while on the lake. Thankfully, it didn't happen.

Once back on the mainland, I tried to figure out which of the remaining three islands had what I was looking for. There had to be an answer. There always has been one up till now. The two writings were the only pieces of information I had since landing here. One was the riddle about the parents of satellites, and the other was a pattern.

Wait a minute! I quickly returned to the overlook, where I found the plaque. I picked it up off the ground and stood it upright. *Of course! It wasn't a pattern; it was an outline.* Looking closely at the outline on the sign and then the islands in the distance, I saw that one of the island's profiles matched the outline on the sign. The correct island to go to was the one whose profile matched the outline. I hurried back to the boat and crossed the lake once again over to island number three. Again, I was very relieved to have made it without it raining.

Upon arrival, I found another dock to which I tied the boat. This island didn't seem to have any specialties like island one's tall hill. From where I stood, I began looking for areas of protection should it begin to rain again. I saw several places like small caves and sizable cracks in boulders that I could hide in.

Again, a path in front of me led from the dock into the inner parts of the island. As I set out along the path, I came to a boulder at the base of a mesa. To continue, I saw that I would need to climb the mesa and that climbing gear would be needed. The ascent was not very easy due to how difficult it was to penetrate the icy coating on the wall of the mesa. After 20 minutes, I reached the top and looked around while taking a breather.

Noticing where the path continued at the mesa's base, I scanned ahead and saw a small white structure some distance away. Along the path, I could see that it came to a fork. Going left, I would come to a boulder that I could use as protection from the rain. After the boulder, I would reach the edge of a cliff. The left fork is not the way I should go unless it starts to rain before I reach the fork. Going right at the fork would bring me near to what appears to be the island's second-highest peak, a peak with a very sharp point. Beyond the peak is another boulder and then another fork in the path right after it. I also see that going right will eventually bring me downhill to another boulder on the edge of a very deep depression. It wouldn't surprise me if this depression were the island's lowest spot.

Intrigued by the sight of the depression, I decided to take a closer look. Fastening my grappling hook to a sturdy boulder atop the mesa, I repelled down its slippery slope. Once at the mesa's base, I continued walking over to the depression for another ten minutes. The sight of the depression up close was rather unsettling. I felt as though I was looking into the silenced mouth of an ancient creature. I got as close to the edge of the depression as I dared. I saw that the drop-off had to be over 200 feet. I was probably looking into an ancient crater eroded by time and weather. The ground at the rim of the depression was quite unstable. Pieces of loose ice and rock began to crumble and break under me. I backed away before it became too late, not wanting to roll down the steep slope. I, therefore, returned to the fork in the path, continuing to the left, which passed alongside the island's highest peak. This, like the depression, was truly a sight to behold.

Shortly after passing the peak, I came across another boulder. My timing was quite fortunate, seeing that it began to rain moments after passing it. I waited for what seemed to be a longer time than usual. When I felt safe that the rain was over, I continued, reaching another annoying fork in the road. Taking the right fork brought me to a cliff with a beautiful view of what seemed to be another land mass far off in the distance. The view was very striking. The combination of the orange-tinted sky and the methane lake beneath it was very relaxing. The sky showed patches of purple as the distant setting sun feebly illuminated it.

Breaking myself away from this peaceful scene, I returned to take the left path at the fork. Doing so finally brought me to the small structure I had seen earlier from atop the mesa.

As I approached the building, I noticed that it had a large overhang at the door. The door itself had something on it that I couldn't quite make out at my distance. Once at the door, I saw nine circles, each smaller circle nestled inside the others until the largest outer circle. At the top of each circle, I saw a small rectangular opening with a symbol inside each one. I'm no genius, but I'm sure the symbols represented each of the planets. Now, I needed to know the order in which to display the symbols.

Being used to how things have been going, I reviewed the items I had found since my arrival on Titan. The only information left unsolved was the sign, which read:

"Parents of satellites - most to least."

Hoping to continue being correct, I assumed the order was to place the planet with the most satellites first. I would then continue until the planet with the least. Back to the book for satellite information!

"Wow! Saturn currently has at least 285 known satellites," I thought. *"That's definitely the planet with the most. Jupiter is in second place with at least 105 known satellites."*

Looking at the information, I placed the symbols to have Saturn first, followed by Jupiter. Uranus came next with at least 28 known satellites, Neptune with at least 16 known, Pluto with five known, Mars with 2, Earth with 1, Mercury and Venus with 0. I was confused about the proper order for Mercury and Venus, seeing that both have no satellites. I, therefore, positioned them based simply on alphabetical order and hoped for the best. To my great delight, the door unlocked, allowing me to enter the structure.

Inside, there was a small table with a metal box resting on top of it. Seeing this made me smile from ear to ear. As I approached the metal box, I saw an odd area of the floor in the far right corner of the room. After looking at it from the table for a few moments, I focused my attention back on the box and removed the much-desired fragment of paper. I would soon reunite this fragment with its family. Mission accomplished! It was time to head back to the Lotania.

But first, I wanted to check out the odd area of the floor. Examining it closely, I saw that it was actually a trap door. Opening it revealed a passageway. It was too dark to see very far into it. Turning on my flashlight, I cautiously walked into the tunnel and followed it for nearly ten minutes before seeing some natural light. It took a lot for me not to turn back.

At the end of the tunnel, I emerged onto the surface of Titan near the dock. I can't believe that I missed the opening when I first arrived. Discovering it would have made everything a lot easier. But no matter now, I have the paper fragment I came for.

After entering and untying the boat, I began moving toward the mainland. Half way between the island and the mainland, my luck ran out. The boat's engine decided to give up. I checked the obvious first and found that the gas guzzler engine ran out of gas. *"Fantastic! I'm in the middle of a methane lake; it could start raining at any moment, and... oh, forget it!"*

When I thought things couldn't get any worse, I heard thunder in the distance. It didn't take long to realize that the weather disturbance was getting closer to my location, especially when I saw a lightning bolt strike the ground about a half mile away. I felt like a sitting duck with a bull's-eye painted on it. I was the tallest object on Death Lake, as I've since come to call it. No good could come from this situation. If only I had some gas left in the blasted gas can.

"*Wait a minute! What's wrong with me? I'm sitting on a lake of fuel. I have all the gas I need,*" I thought.

After filling the empty container, I quickly filled the boat's engine. Just then, another bolt of lightning struck the ground, hitting the methane lake. *Great!* A fire ball the size of a building erupted where the lightning struck. The blaze spread out rapidly in all directions. Thankfully, the engine accepted the liquid methane and started right away.

I raced for the mainland with a carpet of flame right behind me. The fire was about to overtake me as I ran the boat up on the shore at top speed. The sudden jolt of hitting the land practically threw me out of the boat. Thankfully, the fire did not leave the confines of the lake.

While running from the damaged craft as fast as I could under Titan's conditions, the boat exploded into fiery pieces. I fell to the ground as pieces of the boat came down around me. When the excitement subsided, I arose to my feet and staggered over to the Lotania.

Once inside, I collapsed with exhaustion. Again, what was meant to be a few minutes of rest turned into a nap of close to three hours. When I awoke, I removed my suit and boots, stored my new paper fragment, and then slowly made my way over to the command chair. SAROS gave me her normal line about imputing another coordinate, which I did. I entered the number 7872039, which would take me to Uranus' satellite named Miranda. I pressed the engage button, and off I went.

Chapter 13
Miranda's Overlook

The distance between Saturn and Uranus made my next destination last about seven minutes. The ship had to traverse nearly a billion miles of space. Approaching the planets to get to their satellites has been amazing. I can't possibly put into words the majesty of seeing the worlds up close.

About a minute out, SAROS spoke the following:

"Your destination is Miranda, Uranus' fifth satellite. The average distance from Uranus is 80,399 miles. Its diameter is 293 miles. Miranda's length of day is 1.413 days. The length of a year is also 1.413 days. The average temperature is -351.7 degrees Fahrenheit. The atmospheric gases are essentially a vacuum. The surface gravity measures 0.79 % of Earth's. The Lotania will land atop Verona Rupes, the solar system's tallest cliff. For other important information, please read the planetary book."

As the ship descended, I noticed that this moon looked like someone patched it together using super glue. Its features differed from one area to the next. In one location, I saw a huge backward check mark, which I later learned is called the "Chevron." Further south was a feature that was bright white. As the ship continued to descend, I saw that the white area was where the Lotania would be landing. SAROS called the location Verona Rupes, the solar system's tallest cliff. Another location worth mentioning as the ship descended were fifteen circular areas that seemed too strange, even for this place. The circles appeared to be man-made, or should I say, alien-made.

The Lotania fired its cables into the ground and touched down on the surface. I prepared myself for a float on Miranda. To keep from possibly floating away, I tied one end of a rope around the ship's outer rail and the other around myself before heading off.

Making my way over to the mysterious circles was a difficult task. Looking at the circular areas more closely proved that my idea of them being made by hand was correct. Two of the areas were simply holes; another was damaged. The damaged area seemed to be a covering of some type. Peering into the damaged area showed another seemingly deep hole. I had no idea what the area was, or how far down it descended.

Looking at my surroundings, I saw a stark but beautiful landscape. Like before, the stars were indescribable. Bouncing carefully, I headed toward the cliff's edge. Not getting to close, I looked over the edge to see a breathtaking view of the landscape beneath me. It wasn't easy making out anything in detail due to the distance the cliff dropped. I read later that the cliff measures about twelve miles high. I never thought I was afraid of heights until I came face to face with the volcanic shaft of Olympus Mons and now the majestic Verona Rupes. Backing away, I continued my bounce and came across a signpost leaning on an angle in the icy surface. Although completely odd, I was used to seeing messages in strange places, and therefore, didn't give it a second thought. On the sign was written:

"Focus the light to reveal your path."

"Another riddle to solve! Just what I needed!" I thought. *"Just once, I'd love to land somewhere and find the paper fragment just sitting there. I won't hold my breath waiting for that."*

After the sign, I found nothing more. Not certain how to proceed, I thought over my options. I didn't have many; in fact, I only had two. I could return to the ship and try to land somewhere else, but I'd have no idea where. Trying to spot a small metal box while flying a ship manually did not sound like the answer. Besides, the Dearwigs always had a reason for where the ship landed up till now.

This only left option 2: repel down the twelve-mile-high cliff in the dark. I can't say that option 2 thrilled me at all, but that's a personal problem I'd need to deal with. I, therefore, looked more carefully over the edge this time and noticed the cliff wall had several ledges jutting out of it. Many of the ledges seemed to offer a good amount of landing space. I, therefore, decided to repel from ledge to ledge. This wasn't going to be as bad as I thought.

Locating the nearest ledge from where I stood atop the cliff, I shot my repelling cord into the ground. The result was a cord that ricocheted off the ice, almost hitting me. The attempt launched me slightly off the ground due to the extremely low gravity. After stabilizing myself, I contemplated how to continue. It seemed that this ice was not going to play nicely. "*Fantastic! Now, what do I do, jump down from ledge to ledge? Or should I say, slip from ledge to ledge? Not happening! Slipping on a ledge or missing it all together would be certain death. Or would it, I thought? The gravity here is very low. Perhaps my descent would not be harmful if I didn't fall straight to the bottom. It's still a big risk, but it's possible."*

Out of curiosity, I took out the planetary handbook to find out the acceleration due to gravity for Miranda. It did not say directly, but being pretty good with math, I calculated a value of 0.0768 m/s^2. On Earth, the acceleration due to gravity is 9.796 m/s^2. This is quite a difference.

If I did jump the entire distance all at once, it would take me eleven minutes and forty-nine seconds to reach the ground. Unfortunately, although it would take that long, I'd still impact the surface at a velocity close to 122 mph. I'd need a nice airbag to land on at that velocity. Unfortunately, I don't remember the ship carrying one of those. Therefore, I'm back to my jumping from ledge to ledge idea. I just hope there are enough ledges and that they're all close to each other.

Taking one last look as best I could in the dim light, it seemed like the ledges were pretty consistent. It almost appeared as if they were purposely placed. They looked rather evenly spaced apart as far down as I could see. I was beginning to become suspicious about all this.

Ready to go, I took my first plunge. I timed my drop to be about one minute and nineteen seconds. This corresponded to a height of just less than eight hundred feet. Although manageable, I felt the impact pretty well when I landed on the ledge. Timing my next few jumps, my results were similar to the first. I was pretty certain now that the ledges were conveniently placed. At least I can take comfort in knowing that going down the cliff wall was expected of me. In total, eighty ledges guided me down the wall of Verona Rupes. My whole descent took about two hours to complete. Looking back up the wall, I couldn't see the top of it. It was a staggering and mind-blowing sight to behold.

As I looked around from the bottom of the cliff, I noticed five posts standing about fifty yards away. I saw them because something sparkled on a few of them. When I got closer, I saw that each post had a circular mirror attached to it. The mirrors were at different heights on the posts; each mirror having two ropes tied to them on opposite ends, reaching almost to the ground. Pulling the ropes, I saw that the mirrors could be spun 360 degrees on their axis. It must have been the distant sun's light reflecting off a few of the mirrors that caught my attention.

Looking at the cliff wall from my new position, I saw several cave openings. They were arranged in levels. There were three openings at the lowest level: five a little higher up, five more above the first group of five, and two more at the highest level. It was odd to have fifteen openings in a cave wall, all within the same area. Come to think of it, there must be a connection with the fifteen circular areas I saw at the top of the cliff. The circular areas are right above the cave openings.

My next move seemed fairly obvious; it was time to search the caves. I wondered if any of the caves were more important, that I might examine them first. Not knowing the answer, I figured I'd search them one at a time. I had a similar problem choosing which island to visit while visiting Titan. At least I had a clue to help me with that problem.

Slapping my head again, I realized I also had a clue here. *"What was it the sign had written on it? Focus the light to reveal your path. Well, that makes a lot more sense now."* Studying the mirror puzzle before me, I tried working out how to use all five mirrors to reflect the Sun's light and reveal my much-desired path.

Starting with what seemed to be the lead mirror, I tried to reflect the Sun's light over to its nearest neighboring mirror. I eventually established a link between the two and began fiddling with the third mirror. This process continued until I successfully had the Sun's light reflecting off all five mirrors. I made a bad mistake when the beam of light struck me unexpectedly while adjusting the mirrors. One of the mirrors reflected onto me from its initial resting position before I could adjust it. I felt the temperature around me increase rapidly. I moved the mirror quickly to stop the beam from hitting me. I understand now how a Thanksgiving turkey feels while in an oven. Not fun!

With all five mirrors reflecting the sun's light, I figured I was successful when I saw a beam of light project away from the fifth mirror to one of the cave openings. The beam shone upon the cave opening on the third level, second from the end on the right side. Knowing where to go, I approached the cliff and looked for hand and foot holds to ascend it. I knew my cable gun would not work due to my failed attempt earlier. Not seeing any safe way to proceed, I made several attempts to reach the cave. All of my efforts were of no value. I slipped off the wall with every attempt, unable to get proper footing. Out of desperation, I tried the cable gun again. I fired as close to the cave entrance as I could. To my astonishment and delight, the cable firmly hooked to a protruding piece of the rock face. Tugging on the cord, satisfied that it would hold, I began going up the cliff wall to the cave.

Standing at the entrance, I could not see anything inside the cave except a small light that appeared to float about three feet above the ground. From where I stood, a huge void could have been waiting to devour me. Not comforted by my thought, I took out my flashlight before entering. Shining the light around, I still didn't see much. But what I did see made all the difference. A beautiful metal box was sitting on a small table along the back wall. Excited, I made my way over to the table. I was moving a bit too fast and slammed into the table and then the wall. After stabilizing myself, I opened the box to see the paper fragment waiting to be taken.

As for the light I saw that seemed to float, it belonged to what looked like an elevator button. It was attached alongside a cylindrical opening that looked like those old tubes used at banks to get your materials to the banker. It was probably not my most brilliant move, but I stepped into the cylindrical space and pressed the lit button. I then found out how the stuff inside the bank tubes would feel if it could. I took off at a pretty decent speed, heading….. somewhere. A little over a minute later, although it seemed much longer, I found myself at the top of Verona Rupes again. I was standing in one of the circular areas I saw when I first arrived. Although, when I first saw this circle, it was just an empty hole. Now I know why.

I stepped off the circular area, thinking how much easier and quicker all this would have been if I knew these were elevators. To be sure, I placed a descent-sized rock on one of the other circles, and down it went. I stood there staring at the void where the rock was, feeling cheated and foolish.

Breaking free of my stare, I slowly headed back to the Lotania. Of course, if I had stepped on a random circle, I wouldn't have known the correct cave to enter. I may not have been able to travel from one cave to another. Having thought all this, I convinced myself that the action I took to descend the cliff was the correct action. Feeling a little better, I entered the Lotania. I put away my clothing and paper fragment, and then returned to the command chair. SAROS requested another code. The next on the list was called Triton. I input the coordinate 8841426 and pressed the engage button. My adventure continues.

CHAPTER 14
TRITON'S SENTINEL OF ICE

Approaching Neptune was striking, just like his gas giant brothers before him. The deep blue appearance of this frigid orb will never be forgotten. As the Lotania zeroed in on Triton, I noticed major differences in its landscape. One half looked rough, like the skin of a cantaloupe. The other half appeared much smoother. I can say that I never had much appreciation for the solar system before this adventure, but now everything has changed. The beauty and mystery that exists so far from home is unreal, yet….very real.

As the ship began to descend, SAROS spoke the following.

"Your destination is Triton, Neptune's largest satellite. The average distance from Neptune is 220,437 miles. The diameter is 1,681.9 miles. Triton's length of day is 5.877 Earth days. Its length of year is also 5.877 Earth days, only retrograde. The average surface temperature measures - 391.3 degrees Fahrenheit. Triton's atmospheric gases are traces of nitrogen and methane, and the surface gravity measures 7.95 % of Earth's. For other important information, please read the planetary book."

As the craft continued to descend, I pondered on the fact that Triton's orbit around Neptune was retrograde, which meant that Triton would eventually spiral itself into Neptune sometime in the future. I was trying to imagine such an event when the Lotania touched down on Triton's surface. Breaking free of my thoughts, I got dressed again in the same suit and shoes I just took off a few minutes earlier. I should have just stayed dressed. I may double-check the conditions of the next place before taking the suit off again.

As I opened the main hatch, I did it with confidence, as though my actions were a normal every day activity. After exiting the craft, the first sight to catch my eye was a monumental frozen geyser that stood hundreds, if not over a thousand feet. It was a beautiful ice-covered structure that gleamed in the distant sunlight. It appeared clean and clear like water had recently poured down its impressive slopes. Next to it was a cliff that actually stood a little bit taller.

Built alongside the geyser was a crude-looking lift that functioned by a system of pulleys. Two large basket-shaped carriers were attached to the device, one at the bottom and the other at the top. I figured that the basket on the bottom was what I'd be using to get to the geyser's top. All around the crater of the geyser was a catwalk; at one point in its path, an attachment led right over the crater itself. Sooner or later, I knew I'd have to go up there. I hoped it would be later, much later.

To get a better look at the elevator-like mechanism, I approached the geyser and found a small metal crate at its base. Inside, I found three charges of C-4 explosive, each labeled with a different frequency. The frequencies were 270, 400, and 520 Hertz. Alongside the explosives was a remote-controlled detonator with a dial that could be set to the frequencies just mentioned. Lastly, the crate also contained a sextant for making measurements. It's been a long time since I've seen one of these beauties.

Continuing to look around, I took an interest in the cliff that rivaled the geyser's height. The two geologic sentinels neighbored each other, making an impressive sight to behold.

Another object that needed to be examined was at the base of the cliff. Glancing down at the crate of goodies before leaving, I noticed something sketched on the underside of the lid. It showed a drawing of a person facing the Sun, using the sextant. I understood what the drawing was showing; I just didn't know why it was showing it.

As I approached the base of the cliff, I could see that the mechanism took on the shape of a see-saw found at a playground. A large and thick metal plank balanced at the device's fulcrum. A close-up investigation of the device revealed detailed markings for altitude measurements ranging from 0 to 90 degrees. An engraving of the Sun was also placed above the degree markings. Located below the markings was a lever in the upright position. Pulling the lever down locked the plank into its present position. I needed to learn the purpose of the device. It was time to go over all that I knew so far.

I had explosives with a detonator, a sextant, a see-saw-like device with degree markings, and a rickety-looking elevator that I was not in any hurry to get on. *"How does it all fit together?"* I thought. I knew I was expected to measure something with the sextant, and I'm sure it had some connection with the see-saw device due to the degree markings on it. *"Ah! Wait! An engraving of the Sun is also found on the device. And yes, the crate's lid had a sketch of a person using a sextant on the Sun. What purpose is there in measuring the Sun's altitude? How does it all connect?"*

I head back to the crate to see the lid and retrieve the sextant. Gathering no further clue from examining the lid again, I focused on the sextant. I aimed the instrument in the direction of the Sun. Lining up the sextant's zero-degree mark with the horizon of Triton. I measured the Sun's altitude to be twenty-two degrees above the horizon. Returning to the base of the cliff, I unlocked the device and set the metal plank to an angle of twenty-two degrees. I then locked the mechanism again. *"Ok, now what?"* I thought.

Looking around to see if I was missing vital information, I saw the elevator again. Looking back and forth between the elevator and the see-saw device, I noticed the plank aligned with the elevator's baskets. *"Of course!" "It's a catapult,"* I thought. I needed to launch something into the basket at the top, which would carry the basket on the bottom upward to the catwalk. The twenty-two-degree angle setting would enable me to launch an object at an angle of sixty-eight degrees, which looked like it would be just about right to land something in the upper basket.

I looked around for an object to launch. It didn't take me very long to find something. All around the area were boulders strewn about. To move the boulders, I needed to use the Land Rover. Entering the Lotania and the rover, I backed the vehicle onto Triton's icy surface. As I approached a boulder, the rover shimmied back and forth on the ice.

Didn't I just say that Triton's surface was icy? What was I thinking? I head back to the ship to get the rover's tire chains. Once put on, driving became much easier. After picking up one of the boulders of icy rock, I drove to the see-saw device and placed it on the metal plank's lowered end. The plank had braces on each end to hold an object in place. I'm telling ya! Those Dearsings thought of everything.

Now, I had to launch the projectile. I had to get another boulder, which would need to be dropped off the cliff and land on the metal plank's raised side. I would drop the boulder using the robotic arm.

"No! I can't do that," I thought. I must be in the bottom basket when the boulder hits the plank. *Of course, Nummy! That's what the C-4 and detonator is for.*

Grabbing a charge from the crate, I carefully took it with me as I retrieved another boulder. The drive up the cliffside took some time to complete. The rover occasionally lost traction, even with the chains, sliding backward a little. I was relieved when I finally reached the top. After a few moments, I carefully placed the boulder near the cliff's edge and backed the rover away. I exited the vehicle and placed the C-4 charge directly behind the boulder opposite the cliff's edge. I was set to put my plan in motion.

Driving down the cliff was not as bad as ascending it. I just took it easy and soon was approaching the Lotania again. I exited the rover and walked to the elevator mechanism. I picked up the detonator from the crate; got inside the basket, and prepared to blow up the charge.

"WHOA! Just a second," I thought! *"I can't just start pressing buttons. Instead of blowing up the charge near the boulder, I could blow myself up."*

Gathering myself momentarily, I headed back to the crate to see what frequency charges were still in there; then, I'd know which frequency was parked under the boulder. I was fortunate to have looked; the charge frequency under the boulder was 270 Hertz. The detonator was set to

520 Hertz. Had I brainlessly pressed the button in haste, I would have become a permanent resident ALL OVER the surface of Triton.

Before heading for the basket again, I put the remaining C-4 charges with my gear. Once again, I entered the basket that would *hopefully* take me to the top of the geyser. Correctly setting the detonator to 270 Hertz, I set off the charge. It was an impressive sight as the boulder lurched forward and fell to the ground beneath. What added to the neatness of the event was that there was no sound or fire when the blast happened.

The boulder came down with part of the cliff's edge, taking about half a minute to do so. This made my original guess of the cliff and geyser's height close. As planned, the boulder hit the metal plank. Although the device was locked, the impact launched the ground-based boulder into the air, and then amazingly into the upper basket. This, in turn, carried me up to the catwalk encircling the geyser's crater. *"These Dernigs are incredible to have worked all this out. I can't wait to meet them"* I thought.

The ride up to the catwalk was shaky at best. When the basket came to a stop, I wasted no time in getting out of it. Walking around the path offered a nice view of Triton's surface. The icy terrain sparkled in the distant sunlight as though thousands of diamonds were scattered about. From where I now stood, I could see the metal box sitting on the crater's rim. As I made my way over to it, I felt a sizable tremor that shook the entire catwalk. Hoping it wasn't the geyser getting ready to clear its throat, I looked over the walkway's edge. To my amazement, the boulder that I blew off the nearby cliff was resting on the base of the elevator structure. It crossed the icy landscape and crashed into the elevator, damaging the structure's base.

Moments later, the walkway shifted again. The sounds of metal grinding and ice cracking would have been unnerving if they could have been heard. Ahead of me, the walkway began to collapse. I hurried back to the crater and jumped off the walkway. I watched as the entire catwalk broke apart piece by piece. In minutes, the catwalk lay as a twisted mass of useless metal on the icy surface below me. If circumstances were not so dire, I would have found sitting on the rim of the geyser's crater quite exhilarating. But conditions were nowhere near acceptable. I was stranded up here.

Before forgetting, I opened the metal box and took the paper fragment. I then immediately began looking for a way down. Walking around the rim of the crater was extremely treacherous. I won't tell you how many times I almost slipped.

While scoping out the rim, I felt another tremor.

"How much more damage can that blasted boulder do? The whole structure has been destroyed," I thought.

Looking at the ground, I saw that nothing was moving. I sat there perplexed for a few moments. Then my eyes opened very wide.

"No! It can't be," I thought.

Reluctantly, I slowly looked down the crater. A large column of bubbling water was making its way to the top. In a severe panic, I found the smoothest area of ice the geyser had to offer so I could slide down the side of it. I was not in the mood to become a permanent fixture of ice on this distant moon. I could live with it if I died trying to escape death, except I wouldn't be alive. Oh! You know what I mean.

Choosing my path of descent, I sat down on the crater's rim and pushed off. Despite the low gravity, I still moved at a pretty good velocity when I reached the bottom. The sliding took about thirty-five to forty seconds to complete the trip. I'm sure I was moving over thirty miles per hour when I reached the bottom. I figured this to be true because I traveled quite a distance from the geyser's base before I stopped. I laid there on the ground, a mass of bruises from all the knocks I took on my way down.

Another tremor, stronger than the last one, scared me to my feet. Forgetting about my bruises, at least for now, I returned to the Lotania. I noticed that the rover was still parked outside as I approached the craft. Hobbling over to it, I quickly drove the rover back into the ship.

Once inside, I closed all the doors of the craft. Making my way to the command chair, I input the coordinate 9296718. I did it so fast that SAROS didn't bother to say anything. I hit the engage button, and the ship took off. Looking out the window while the ship ascended, I saw the geyser erupt a beautiful plume of water, which instantly froze as it made contact with the ground. *"That was too close,"* I thought. I needed to erase this near tragedy from my mind as soon as possible.

Chapter 15
The Plutonian Maze

Lost in the thoughts of my last escape, the Lotania made its way to the dwarf planet Pluto. Transferring my thoughts of nearly dying on Triton to the amazing fact of going to Pluto wasn't difficult. I had no desire to continue pondering on disturbing negativity. I was about to land on a planet that man has only taken a decent picture of. As the ship began its descent, like clockwork, SAROS faithfully chimed in with her information about the destination below. She said the following:

"Your destination is the dwarf planet Pluto. The average distance from the Sun is 3,670,052,065 miles. Pluto's diameter measures 1,476.8 miles. Its length of day is 6.405 R days. The length of year is 248.08 years. The average surface temperature measures – 380.5 degrees Fahrenheit. Pluto's atmospheric gases are mainly methane and nitrogen, and the surface gravity measures 6.33 % of Earth's. For other important information, please read the planetary book."

As SAROS spoke, I gazed out the window at the landscape below me. Like recent destinations, I saw a barren and icy surface. Even still, it had its definite uniqueness and beauty. When the ship landed, I took things a bit slower this time. As I mentioned earlier, I wouldn't remove my suit unless necessary. It paid off to leave it on.

Being dressed already, I took a few moments to see how the pieces of paper fragments were coming together. After all, I had ten of the twelve pieces of paper. I was sure they would fit together pretty nicely by now. I was not disappointed when I attempted to piece them together. I was only missing an edge and an internal piece that fit a little left of center. Even though I had most of it, I couldn't make out heads or tails of what it was saying. I only knew what Richard said: that they were directions to a distance race of people in a solar system beyond our own. I still couldn't wrap my head around that. I had no idea how I would read the completed document, but I'd face that challenge later. I had Pluto to deal with right now.

Exiting the craft, I received the odd sensation of near weightlessness again. I showed no real danger of floating off Pluto's surface as long as I walked softly, not attempting any heroic moves. Knowing where to investigate next wasn't too difficult. I safely assumed the cave with the stone box in front of it was the right move.

Nothing was all that easy to see. The Sun appeared very tiny in the starry sky above. The light it gave off was comparable to lighting a room with a small seven-watt bulb. For safety, I used my flashlight during my stay on Pluto. After making my way over to the cave, I removed the lid of the stone box, which offered no real challenge. Without the occurrence of weather on Pluto, the lid was not frozen by any rainfall. Inside, I found another box, this one made of metal. *"No! That would be too good to be true and way too easy,"* I thought.

Being very hopeful, I looked inside the metal box and found a piece of paper. *I can't believe it! Are the Dearwhegs cutting me a break at last?* Shining my flashlight on the paper revealed the following:

If P = 8, then X is double the value of P.

Know, therefore, that F = (P + X). Also, know that L is 1/4 of X.

Now B = (L + F) - P. Therefore, T = (B - P).

With all this being said, let K = (T / L) and N = (F / L).

Let J = [(B + T) / X], and R = [((X - N) / J) + (P - K)].

This will make M = (R - M).

Finally, W = [((J^N - (M + P)) / K) + (X + K + B + R) + ((T + F) / L)] / M.

Finally, let I = X - W.

(The value of "I" breaks the code)

I must warn you that I give the answers to this puzzle as I continue explaining my adventure. If you wish to solve it on your own, do it now. Good luck!

"So much for the Dernwacs cutting me a break! I knew it was too good to be true. And what is this mathematical gibberish?"

Enjoying math and being curious, I tried to decipher the writing right away. The answers I came up with were P = 8, X = 16, F = 24, L = 4, B = 20, T = 12, K = 3, N = 6, J = 2, R = 10, M = 5, W = 15, and I = 1. I looked forward to knowing what all this meant.

As for the cave, it almost looked normal. A flashlight was needed within the cave to see anything at all. The walls and ceiling were roughly hewn rock and ice. Its major oddity was that it was a maze. I found that out by moving around within it for who knows how long. As I walked its corridors, I was faced with many choices on which direction I wanted to go. I ran into several dead ends. In one area, the ground was very weak and collapsed beneath me. I found myself hanging on to the edge of the floor for dear life. Thankfully, I was able to pull myself back up. I almost lost my flashlight down the hole when I fell. Aiming the flashlight beam into the hole showed me nothing but blackness. Dropping a rock into the hole produced no sound to indicate a bottom. *Then again, what else could I expect from an airless world? Of course, there would be no sound. I'm exhausted!*

Physically and mentally shaken, I backed away from the abyss. Before leaving, I examined what the weak area of ground looked like in case there would be more. At one point, traversing the maze, I came to what looked like an exit. Just outside, lying on the ground, I found three sticks of TNT and a detonator. This was all pretty weird: a mathematical puzzle, explosives, and a man-eating maze. Pluto ranked in the top three of the most confusing places.

Look! Going into detail on the entire maze would put you to sleep. I'm sure you'd rather watch paint dry on a wall. So, let it be sufficient to say that I eventually found the end of the maze; and yes, there were more areas of weak ground. I saw at least seven of them. I don't know how many the maze contained overall.

By the process of elimination, anyone walking the maze would find the end of it. There appeared to be only one real way to go. Walking through the maze eventually brought me to a door. On the door was written the word OPEN. Above each letter was a combination dial, each numbered zero to twenty-six. For kicks and giggles, I tried to open the door, and IT OPENED.

Just kidding! It was locked. I probably would have dropped in amazement had it really opened. The fact that the dials went up to twenty-six was a tremendous clue as to what needed to be done. Placing the dials on the numbers 15, 16, 5, and 14 to represent each letter of the word OPEN did absolutely NOTHING. No words could adequately describe the astonishment I had on my face, so I won't even try. *Now, what do I do?* So much for thinking that I knew what needed to be done!

I don't know how long I sat there puzzled. I was getting so hungry; my backbone felt like it was shaking hands with my belly button. I felt like I was going to starve to death. I continued by reviewing the only thing I had: the mathematical note. I just couldn't see what I was missing. I went over all of my calculations to see if I made any errors, and I could find none. I didn't know how the letter "I" could break the code. What did it mean?

It took longer than I care to confess, but I finally figured it out. I asked myself,

"Why I, and not any of the other letters?"

"The value of I came out to equal 1."

"Of course! I = 1!"

I originally set the dials as if the letter A equaled 1. The letter I was being given the status of the new first letter of the alphabet. Therefore, if I = 1, then J = 2, and so on. Coincidently, all the letters come out to equal the values figured out.

Now, using the new alphabet arrangement, I reset the dials to 7, 8, 23, and 6, respectively, and tried the door. To my great relief, the door really opened this time. *I wasn't going to starve to death after all. Wouldn't that be something, though? After all the challenges I've dealt with so far, to be taken out by food deprivation!*

There goes my mind again, off on a tangent. I did warn you, though.

Alright! Let's focus!

Looking beyond the door was a small room with the same roughly hewn walls of rock and ice. In the room was a table with a metal box that contained the paper fragment I'd been looking for. I stored the piece of map with my other gear. I then noticed on the left wall of the room that there were more numbers and three deep, cylindrical indentations arranged in a triangular pattern. The numbers on the wall were 20, 4, 19, 11, and 12, followed by a small spacing, continuing with 26, 23, 10, and 23. Deciphering this message, I got BLAST HERE. *Of course! The holes were for the sticks of TNT I picked up outside the maze's back door.*

Carefully placing the TNT in the holes and hooking up the detonator to them, I was ready to blow down the wall. To be on the safe side, I stepped outside the room before setting off the charge.

It was a pretty powerful explosion from the look of it. I'm surprised the ceiling didn't come down. Thankful that it didn't, I cautiously re-entered the room after the dust cleared. Looking through the hole I just made in the wall, I saw the entrance to the cave. I was also able to see the Lotania in the distance, waiting to take me to another location. Again, I was left to think of how easy it would have been if only I knew my final destination was right behind a wall in front of me.

Glad to leave the Pluto maze, I headed back to the ship. Once inside of it, I checked to see what conditions my next destination had to offer. Being very similar to Pluto's, I kept my suit on once again. Entering the command chair, SAROS asked for a new heading. I input 9856021 for Charon and pressed the engage button. I had just one more paper fragment to go.

CHAPTER 16
CHARON'S PATTERN

The trip to Charon only took seconds. The ship landed before SAROS' information about Charon was completed. While on Pluto, I could see Charon quite nicely due to its close proximity. SAROS' information about Charon was as follows:

"Your destination is Charon, Pluto's largest satellite. The average distance from Pluto is 12,174 miles. Charon's diameter is 753.1 miles. Its length of day measures 6.3872 Earth days. The length of its year is also 6.3872 Earth days. The average surface temperature measures -364 degrees Fahrenheit. Charon's atmospheric gases are essentially a vacuum, and the surface gravity is 2.94 % of Earth's. For other important information, please read the planetary book."

Another amazing place!

Somewhere, not far away, was my final paper fragment. Being anxious to find it, I wasted no more time on the craft. After stepping onto the icy surface of Charon, my feet slipped out from under me. I found myself flat on my back, gazing out into space.

"*Wow! Pluto looks just as incredible from Charon as Charon looked from Pluto,*" I thought.

Rising slowly to my feet, I learned that my cleats would have very little effect on the ice. I needed to be super careful. I'm surprised I didn't slip and fall on Pluto as well. Charon's landscape was similar to Pluto's in many ways. The attention grabber on Charon was a cliff wall that stood about 80 feet in height. At the base of the cliff was a cave.

"*I hope there isn't another maze in there,*" I thought.

Other than the cave, searching the immediate area showed nothing of interest. While walking to the cave entrance, I fell again, almost sliding into a nearby crater. That would not have been good. In fact, it would have been deadly. Considering how hard the ice was, chances are I would never have been able to get out.

Finally making it to the cave's entrance, I saw a great deal of color within. All over the walls and floor were what looked to be self-luminous gemstones. The inside of the cave looked like a large holiday display. The gemstones gave off plenty of light within the cave that I had little use for the flashlight. The stones on each wall and the floor were arranged in an eight-by-nine pattern. The gemstones displayed the colors of the visible spectrum.

Other than the gemstones, I saw a large inclined wall of icy rock along the back of the cave. I also saw two pillars of icy rock. One pillar was in the center of the cave, and the other toward the back right of the cave. The pillar in the center of the cave measured about 10 feet in height. At the top of this pillar rested a large, many-faceted clear gemstone. I could only imagine what its value came to.

The other pillar's height looked to be on the order of at least 30 feet from the cave floor. The pillars themselves were decorated with gemstones but displayed no particular pattern like the walls and floor. At the top of the 30-foot pillar, I could tell that there was an object, but from my distance, I could not make out what it was. *"It looks like I'm going up there, unfortunately,"* I thought.

Examining the pillar up close, I was happy to see it had several natural hand and footholds. My climb was slow and annoying, and I won't tell you how many times I slipped and had to start over. Eventually, after about twenty minutes, I succeeded in my thirty-foot climb.

At the top, I saw a pedestal of stone that stood about three feet tall. The top of the pedestal was flat, measuring around 1.5 to 2 feet square. Within the top of the stone table were twenty indentations, each about one-half inch in depth. The indentations were arranged in a four-by-five pattern. It didn't take a genius to know what needed to go on the table. The question was how they needed to go on the table. I didn't figure I'd be able to randomly place 20 gemstones however I wanted. Somewhere, there had to be information on the pattern the gemstones needed to be arranged.

From atop the pillar, I tried to piece together any clue I could. I didn't have anything this time. There was no written message to figure out, or anything else for that matter. All I had were the gemstones and pillars. Oh yes, also the clear, large-faceted orb. I figured the answer had to deal with the colorful gemstones.

Looking them over in more depth, I noticed again that the gemstones were displayed in an eight-by-nine pattern. On each wall were two sets of the eight-by-nine patterns. The floor had only one pattern near the pillar at the center of the room.

I asked myself, *"Do the nine sets of 72 gemstones have anything in common?"*

The flat pedestal didn't want 72 gemstones; only twenty were needed. I painstakingly studied and studied the sets of 72 gemstones from my perch for quite some time. I was beginning to get a small headache due to my efforts. My lack of food and rest were no doubt contributing to my problem.

Finally, I recognized that there were twenty gemstones within each eight-by-nine pattern that were the same. An identical color scheme using twenty gemstones occurred on all nine patterns. That was a tough challenge!

Knowing the answer to the gemstone puzzle, I carefully climbed down the pillar. Alright, I fell down off the pillar. It was an odd sensation again, taking a little over 7 seconds to hit the ground. Also, I failed to mention that along with the pattern on the floor; there were a large number of loose gemstones scattered about. I didn't think it relevant to say anything about them. But now I see that I am in need of them.

Once I got to my feet, I gathered up the twenty correctly colored gemstones needed for the gemstone puzzle. Carrying them in my backpack, I slowly made my way back up the 30-foot pillar. I must have been improving in my climbing; it only took me 12 minutes this time, and I fell only once.

Standing in front of the stone table again, I arranged the gemstones within their indentations to match the repeated pattern on the walls and floor. When completed, a beam of light radiated out from each of the four-by-five gemstone patterns found on the walls. The eight beams converged on the large clear gemstone at the center of the cave. This action caused the clear gemstone to project a huge beam of light straight up to the cave's ceiling, which until this time was in darkness.

With the ceiling now illuminated, I was able to see three important things. The first was a very skinny walkway made of ice about 30 feet above the 30-foot-high pillar. The second was a metal box resting on a small icy platform at the end of the icy walkway. And lastly, the beam of light from the crystalline orb was slowly melting the icy walkway 60 feet above the cave floor. I had to move quickly!

With no other visible way to reach the walkway, I pulled out my grappling hook gun in desperation and fired. The hook latched itself to the walkway by wrapping around it a few times. This made it secure enough to give me a free ride-up by retracting the cord.

Climbing the final amount onto the walkway, I slowly and carefully made my way over to the metal box. This was incredibly difficult. Climbing the pillar seemed like a walk in the park in comparison. Looking inside the box, I beheld my final paper fragment.

I did it! I had all 12 of them now.

With a smile on my face, I placed the paper fragment in my backpack and began making my way back to the grappling hook gun to retrieve it.

"Oh no! What was I thinking? I retracted the cord that would have lowered me down. Well, no matter! I could jump if worse comes to worse, although I'd rather not."

While looking around for another way down to avoid jumping, I saw the ice crack and break due to melting. The area closest to the beam gave way. When it fell, it took my gun with it. On the other end of the walkway, farthest from the beam of light, I saw the inclined wall again. *"It would make a perfect slide all the way to the bottom,"* I thought. I almost fell off the walkway while making my way over to the inclined wall, but I managed to keep my stability. Once at the wall, I stepped off the walkway and dropped to the wall. The separation was only about 3 feet or so. The ride down took about 15 to 20 seconds but was incident-free.

At the end of the ride, I retrieved my gun, which now lay on the cave floor. Exiting the cave, I slowly made my way back to the Lotania. Although the walk back was slow, I did it without falling down. I looked at this as an accomplishment.

Once aboard, I felt it safe to remove my suit and shoes. After relaxing for a few moments, I took out all of the paper fragments collected across the solar system. I placed them on the floor to finally arrange them. *"I've been waiting to do this for a long time,"* I thought.

Still having no idea how to decipher the message, I asked SAROS the following,

"SAROS, do you have any idea what this message says?"

SAROS responded, "Congratulations, Captain! You have collected all of the paper fragments that will lead you to Marathrax."

"WHAT!"

Just then, the ship began to lift off.

"WHOA! WHOA! What are you doing?"

Receiving no response, I ran over to the command chair. A new coordinate of 5473889 had been input, presumably by SAROS.

"Where are we going?"

"To get you the answer to the first question you asked me."

CHAPTER 17
THE IONIAN DECIPHER

Feeling like a confused, right-seat hostage, I anxiously watched as the ship raced somewhere. The coordinate was not one with a known destination, at least not by me. It didn't take long for me to see that I was heading toward Jupiter again. Not long after, SAROS said the following:

"Your destination is Io, Jupiter's volcanic satellite. The average distance from Jupiter is 262,032 miles. Io's diameter measures 2,263.8 miles. Its length of day is 1.769 days. The length of its year is also 1.769 days. The average surface temperature is -261.7 degrees Fahrenheit. Io's atmospheric gas is mainly Sulfur dioxide but essentially a vacuum. The surface gravity for Io measures 18.34 % of Earth's. Be aware that temperatures can exceed 600 degrees Fahrenheit near the lava flows. For other important information, please read the planetary book."

"LAVA FLOWS!" "That's just dandy!"

As the ship descended, I continued,

"SAROS, why have you taken us to Io?"

"You asked me if I had any idea what the message says; your answer lies at the base of the volcano, Prometheus," SAROS replied.

Looking out over the Ionian landscape, I saw the volcano SAROS referred to. At its base was a metal box, presumably holding the answer to my earlier question. The location didn't look at all friendly. There were areas of yellow sulfur surrounding a huge lava flow. There were also bright sulfur dioxide volcanic plume deposits that were circular in shape. If that wasn't bad enough, the volcanic caldera was emitting a large amount of steam.

I foolishly asked SAROS,

"Is this volcano active?"

"Very active, Captain. In fact, my readout of the amount of gas and steam exiting the caldera indicates that Prometheus will erupt again very soon. You may wish to hurry."

"How soon are we talking here?"

"That would be difficult, to be precise, Captain. It could be days, possibly hours."

With widened eyes and a look of total panic, I wasted no more time. I quickly dressed myself for a walk on Io. This time, my suit needed to protect me from high temperatures due to the lava despite Io's normal average temperature. I did not necessarily need cleats, but I wore them anyway. Once outside, I saw that the Lotania sat atop a mesa. From what I could see, it looked like a fairly straight shot over to the metal box. Excited to see what was in the box, I hurried toward the base of Prometheus, named after the Greek god of fire. Not very comforting, I'm afraid!

Getting down the mesa was not difficult at all. While hurrying, I slipped and rolled down its bumpy, but moderate slope. Therefore, no climbing gear was necessary. I made it over to the metal box in about a minute and a half with multiple bruises, but thankfully, no broken bones. While limping my way over to the box, I felt a small tremor, which I chose to ignore with gritted teeth. I will say though that I quickened my pace quite a bit.

Standing at the base of Prometheus, I looked down at a locked box. Kneeling next to it, I saw that the box had a combination lock consisting of six wheels set vertically into the lid. Each of the combination wheels was numbered zero through nine. While still examining the box, the ground rumbled violently. Looking up, I saw lava beginning to flow slowly down the volcano's sides.

I yelled out,

"DAYS! HOURS! TRY MINUTES, SAROS!"

Oh, I guess I couldn't be angry; she did say it would be difficult to be precise. But please, that far off!

With no time to figure out the combination, I grabbed the box and quickly returned to the Lotania. Looking back, I could see the lava coming out much quicker now. It was already at the volcano's base as I began my ascent back up the mesa. Once on top again, I looked back again to see the lava more than halfway to the mesa. Quickly, I entered the ship and closed the doors. I prepared myself for a quick getaway should it be necessary. The lava made it to the base of the mesa and came no closer to the ship. Prometheus was done for the moment. The eruption was not violent but more like a burp.

I needed to figure out how to open the box before Prometheus succeeded in clearing its throat. While I worked on the box, SAROS kept tabs on the volcano, ready to sound the alarm if necessary. Again, I found myself racking my brain, trying to figure out what number would satisfy the box. As far as I could remember, I used all the clues I'd been given up till now.

Going over my adventure, I tried to remember as much detail as I could. I didn't make it too far down memory lane when I remembered the sign on Mercury that had a six-digit number on it with no explanation.

"Of course! I also saw the same number in my mailbox at TransQuest. And wait! It's also right here in the ship on the far left of the control panel." Quickly getting up and almost stumbling on my way to the control panel, I saw at the far left the number 226912. This had to be the coolest moment so far. To think, a number presented to me in three different locations at the beginning of my adventure would be needed to unlock a great mystery now. Well Done Dearweens!

Dialing in the six-digit mysterious number did the trick. Hearing the click of success was wonderful to my ears. Inside the box was the deciphering key needed to decode the message. The Dhirniy alphabet was written alongside the equivalent English alphabet. This should be quite easy now.

"Wait a minute! Did I just say Dhirniy? It just came to me now, after all this time. I've really been butchering their name, something terrible! Well, better late than never, I guess."

Snapping free of my thoughts, I took a quick peek outside to see where the lava was. I saw that it made no further advancement up the mesa. Feeling more at ease, I got to work on translating the message which took me a little over an hour. The final result read as follows:

"Congratulations, my friend! I knew you would be successful. It is time for you to complete your journey. Input the coordinate 9522690912, and set the ship's velocity to 99. 48 % C. I await your arrival."

I found myself staring at the message. The moment had finally come. Although I had worked toward this achievement for some time, I felt as though I had been glued to the floor. Eventually, I found myself moving very slowly toward the command chair. I felt as though I couldn't reason anything out. It was as if my mind were lost in a dense fog.

When I finally sat down in the chair, SAROS' familiar request for a new coordinate was asked of me. Her words startled me as though I was hearing them for the first time. I guess I was pretty nervous, even to the point of being reluctant to proceed with my instructions despite all I had done to get here.

"I was being silly," I thought. *"Of course, I had to proceed. I couldn't turn back now after coming this far."*

I, therefore, input the ten-digit coordinate, noticing that I was grinding my teeth while doing it. After the last digit was input, SAROS said, "Please set the velocity at which the Lotania will travel and engage the craft."

I could feel a cold sweat all over my body. I felt as though some of Prometheus' lava made it under my chair after all. Nonetheless, I proceeded to set the velocity as instructed and then pressed the engage button. When the ship's engines revved, every nerve in my body tensed.

Just then, the engines powered back down again, shutting themselves off completely. With my mouth opened slightly, I found myself staring at the only thing still active on the panel in front of me: a lit manual control button.

This page is for the Dhirniy alphabet picture.

This page is for the Dhirniy numbers picture.

Part 2
The Marathrax Incident

CHAPTER 18
WORLDS BEYOND

"What is this, SAROS? Are you kidding me or something? Is this some kind of a joke?"

"I'm afraid I don't understand, Captain."

"What do you mean you don't understand? I'm staring at a manual control button that is waiting to be pushed."

"Then you should push it, Captain."

"Are you trying to be funny? I'm not ready to fly this craft on my own, especially if we need to use relativistic speed."

"You will be fine, Captain. Just remember to keep the ship level whenever you take off or land."

"I appreciate the vote of confidence, SAROS, but this is nuts."

"Captain, you are going to be fine, and remember, Mr. Gaber is waiting for you. Oh, and one more thing. The relativistic jump to 99.48 % the speed of light will automatically take place thirty seconds after the ship is engaged. It would be wise to be level at this time, and be as far away from Io as you can get."

"Any other good news, SAROS?"

"No, Captain, just good luck, and enjoy the trip."

"I can't believe this is happening," I thought.

Ok! Let's just take this nice and slow.

Taking a deep breath, I pressed the manual control button. The ship came to life again, lighting up like a Christmas tree. SAROS then said,

"All systems ready for departure, Captain."

Tensing this time while taking another deep breath, I pressed the engage button. The engines revved again, the thrusters fired, and I practically had to change my pants. SAROS began the 30-second countdown. I gripped the ship's joystick so hard, I was surprised I didn't leave an indentation on it. The ship lifted off of Io, and I kept my eyes on the level indicator.

"So far, so good!" I thought. I then asked,

"SAROS, is there any way to make the ship ascend faster?"

"No, Captain," replied SAROS. "It won't be necessary. Just keep it level, and we will be plenty distant from Io's surface. Nineteen seconds until the jump to near light speed. We will be in orbit around Io in 12.3 seconds."

As the countdown continued, I kept the ship level. I have to admit, I thought it would be more difficult than it was. SAROS then spoke,

"We are now in orbit around Io, Captain. Six, five, four, three, two, one…the drive is engaged."

"WHOAAAAAAAAAAA! TRIP AND TRIP MAN!"

There were no words to describe what I was seeing. The view outside was wild. All the stars seemed to fold in toward the front of the ship. I was able to see what was going on behind me but out of the front view screen. I know it doesn't sound like it makes any sense, but neither did what I was looking at.

The relativistic jump lasted for about 20 seconds, and let me tell you, I would do it all over again just to experience another jump. WHAT A RUSH!

When the drive disengaged, the view of the stars gradually returned to normal, and the ship returned to space normal velocity. Despite the wild view being over, I was by no means disappointed at what I saw next. Before me was the awe-inspiring sight of a giant orange star, much bigger than our Sun. From my distance, I could see that it had planets orbiting it.

I asked SAROS, the following,

"SAROS, is there anything you can tell me about the solar system ahead of us?"

"Yes, Captain. We are 370.2 light years from Earth. The giant orange star seen ahead has a K3III spectral class, with a temperature of 4,332 Kelvin. As seen from Earth, the star has a magnitude of 4.97. Sensors show that the star is 234.88 times more luminous than the Sun and 27.26 times larger. The star has been given the name Uganthos. It is known as 42 Librae on Earth."

I was so amazed with everything before me, that I'd be surprised if I caught half of what SAROS reported.

SAROS continued, "There are seven planets in the solar system seen ahead. The planet closest to Uganthos has a barren and cratered surface, with an average temperature of 282 degrees Fahrenheit. It orbits Uganthos at a distance of 213,788,347 miles, or 2.2999 astronomical units.

The second planet, is the solar system's second smallest, having an average surface temperature of 107 degrees Fahrenheit. It orbits Uganthos at a distance of 366,222,491 miles, or 3.9397 A.U. Its surface is mostly desert, having only an 8% surface area that is covered with water, and it is inhabited."

I quickly interrupted SAROS.

"Hold on! Did you say the planet is inhabited?"

"Yes, Captain." "After scanning the planet, I estimate a population of nearly 1.29 billion."

Utterly amazed, I silently sat there, trying to take everything in. SAROS went on.

"May I continue with my report, Captain, or do you have something more you wish to know concerning planet 2?"

Shaken from my thoughts, I replied, "Ah, Yes, SAROS, you may continue. I may have further questions to ask you about planet 2 at a later time."

"Very good, Captain," SAROS responded. "Continuing on, the third planet orbits Uganthos at a distance of 402,291,549 miles, or 4.3278 A.U. After scanning the third planet, I see that it is practically the opposite of planet 2. It is the second largest of the planets, with an average temperature of 72 degrees Fahrenheit and a surface that is mainly water. In fact, 97% of the planet's surface is water. The only land, making up the remaining 3%, is located at global coordinates, 138.226 degrees east longitude and 26.912 degrees south latitude. I also seem to be reading a structure of some type that I can't quite discern fully at this distance due to distortions."

"Distortions? From where, and what type?"

"They seem to be emanating from the fourth planet, Captain. It would appear that the distortions are affecting time."

"Time!"

I paused to think for a moment and then continued, "Very well, SAROS, please continue."

"Yes, Captain. The fourth planet in the system is the most Earth-like of the planets. It is the Dhirniy home world, Marathrax. It orbits Uganthos at a distance of 440,533,525 miles, or 4.7392 A.U. The average surface temperature is 57 degrees Fahrenheit, and it experiences seasonal changes similar to those on Earth. It is an inhabited world with an estimated population of nearly 1.73 billion. I am also detecting a greater level of technological advancement on planet 4 than what I did on planet 2. Planet 4 also has one natural satellite called Varyn."

“Amazing,” I said. “Two inhabited planets in the same solar system. Oh! Please forgive me for interrupting you again, SAROS; please continue.”

“That is quite alright, Captain; it is all very interesting. Continuing on,.....the fifth planet orbits at a distance of 563,897,917 miles, or 6.0663 A.U. It has an average surface temperature of -3 degrees Fahrenheit. The atmosphere is comprised mainly of Nitrogen, Argon, and Methane. The amount of vegetation on planet 5 is............ Captain, I am reading more time distortions emanating from planet 4. The readings I am receiving measure 30 orders of magnitude greater than the first readings. I am also picking up a weapon’s signature within the distortion.”

“Weapons fire!” “What is going on down there?” I asked.

“SAROS, can the ship be turned back over to automatic control? If so, please do so.”

“Yes, Captain.” “Changing over now.”

“Please set a course for planet 4 and engage.”

“Right away, Captain.”

As the Lotania headed for Marathrax, I caught sight of four objects heading away from the planet. I then asked:

“SAROS, can you identify the objects moving away from the 4th planet?”

“Scanning now, Captain! .... Captain, they are scout ships from the 2nd planet. One of the ships is veering off on an intercept course with the Lotania.”

“Red Alert!” cried SAROS.

“Vessel on radar bearing 346.7 degrees! It is recommended that you raise the Lotania’s shields.”

Quickly, I raised the shields as SAROS suggested. On the view screen, I saw a crosshair image. SAROS then continued,

“Prepare for battle, Captain! All weapons online!”

Not soon after the shields were raised, the Lotania was hit by enemy weapon’s fire. With the ship on automatic, SAROS did some pretty fancy driving to avoid further attacks. Staying close to the alien craft, SAROS maneuvered the Lotania, placing our opponent within the center of the crosshair. SAROS then said,

“Fire, Captain!”

Without much hesitation, I fired a small volley of torpedoes at the enemy craft. It was a direct hit that caused the destruction of the vessel. Nothing drifted in space to indicate that there had ever been a ship. It was totally vaporized. SAROS continued,

"Nice shooting, Captain! Scanning for other vessels in the area!"

After a few seconds had past, SAROS continued,

"All clear, Captain! The other ships have returned to the 2nd planet."

"SAROS, what about the status of the time distortion readings?"

"Scanning, Captain"...................................

"Well, SAROS!"

"Captain, I am no longer reading any distortions..................and I'm also no longer picking up any life signs on the surface of planet 4."

"WHAT!"

I sat there horrified as SAROS stayed silent. *What could have possibly happened to eliminate a population of 1.73 billion life forms,* I thought. After several moments, I then asked,

"SAROS, are you certain that the distortions didn't affect your readings in some way?"

"I am quite certain, Captain," SAROS replied.

In a state of bewilderment, I thought to myself, "*What have I gotten myself into? An entire race of almost 2 billion people have just been blotted out. On top of that, I came out here to establish a friendly relationship with that new race of people, and the first thing I do is kill someone. Certainly, the occupant of the other craft would have killed me had I not defended myself. But still!"*..................

"SAROS," I said aloud, "I need a few minutes to calm down before going to Marathrax. You mentioned a structure of some type on the third planet that you couldn't identify from space. Please allow for a small diversion and head for planet 3."

"Of course, Captain! I understand! Changing course now!"

As the ship approached the third planet, I saw a deep blue world whose color was very similar to that of Neptune's. Just the sight alone brought a feeling of relaxation. It was as though the planet had the power to absorb your very essence in a comfortable embrace. *This is just what I needed.*

As the ship descended, I beheld a surface of water as far as I could see in all directions, except for the small area of land beneath me. It was hard to imagine that the land below me was all the planet had to offer.

Although the water was always coming up on shore, it never did so violently. The very waves themselves seem to practice a form of peacefulness. *"This is truly an amazing place,"* I thought.

On contact with the surface, SAROS said,

"Planet 3, Captain, as requested. The atmosphere is not much different from Earth's. I read a slightly higher oxygen content. No spacesuit will be required. The current temperature is 79 degrees Fahrenheit, with a southeast breeze of 12 to 17 mph. The only thing that may take a little getting used to is the planet's gravity; it measures 10.458 m/s2. You may experience a feeling of heaviness for a small while, but nothing to be alarmed about. Enjoy your rest, Captain."

"Thank you, SAROS, and thank you for your understanding as well. I shouldn't be long."

"Take your time, Captain. I will take readings on the structure in the meantime."

When the Lotania's door opened, a wonderful breeze with a touch of the sea was immediately enjoyed. Oh, how I missed such a pleasure. I may never take breathing air naturally for granted again. Before I began to move, it dawned on me that I was about to step foot on an alien world far beyond my solar system. It was an incredible feeling when I did it on the planets of my solar system, but now, I can't put into words what I was feeling.

Walking off the ship was like stepping onto my own private world. In reality, that's exactly what it was, at least for a short time anyway. Standing there alone, I took in the immense beauty all around me and inhaled deeply as I stretched. Feeling more relaxed than I had in a long time, I collapsed to my knees in the sand and then softly sat on it.

I could see the structure that SAROS detected from space. It looked like a spire with small openings. It rose a number of feet above the water. Truly, it seemed very out of place and shrouded with mystery. I looked forward to hearing what SAROS would have to say about it. Losing myself in a daydream, I laid on the sand looking up into a dark blue sky. The breeze carried me away unto a restful sleep.

I awoke under a slightly brighter sky than the one I fell asleep under. How long I laid there was unknown to me. Feeling a bit lethargic, I stood up and deeply stretched for a good ten seconds. Looking around me, I found myself longing to stay, but I reminded myself that my friend awaited me on Marathrax. Still yawning a great deal, I looked at the water. *"Wouldn't it be nice to take a bath right about now?"* I thought. Since yesterday morning, all I had seen was rock and snow, offering me no chance to take a bath. Taking advantage of this distant blessing, I undressed myself and entered the water. It felt so inviting! Even the water temperature seemed perfect here. Still, the time unfortunately came that I needed to get back to the ship. Unable to sufficiently dry off, I got dressed again with a little bit of difficulty.

As I headed back to the ship, my eyes caught sight of an odd circular patch in the sky that seemed to shimmer. It was barely visible. It was like looking through heat waves while looking through a window on a cold day. I would have passed it off as my eyes playing tricks on me had it not been so circular. Perhaps SAROS could shed some light on the phenomenon. Entering the Lotania, SAROS said,

"Welcome back, Captain! I trust that you had a pleasant sleep."

"Yes, SAROS! In fact, I can't remember the last time I slept so well. How long was I out there?"

"You were asleep for nine hours, eighteen minutes."

"Nine hours!" I exclaimed.

"You must have been extremely tired, Captain, and rightly so after your incredible accomplishments. Are you ready to set course for Marathrax?"

"I thought I might get a little something to eat first and hear what you have learned about the mysterious structure out there. What have you to report?"

"From what I can tell, the structure appears to mark the location of something that resides on the ocean floor. I have taken readings of a strange energy field that penetrates into the depths of the water. The energy seems to originate, oddly enough, from Varyn, Marathrax's natural satellite. From my scans, I can tell you that the structure is hollow, having different levels within it."

I then said to SAROS,

"I wonder if it can be related to the shimmering distortion in the sky."

"Distortion, Captain?"

"Yes, SAROS. Before entering the ship, I caught sight of a circular patch of sky that appeared to be shimmering. I was wondering if my eyes were just playing tricks on me. I was going to ask you about it but became distracted. I'm afraid I'm still a bit lethargic."

"No need to apologize, Captain; it's quite understandable. I'll scan the atmosphere for any anomalies."

A few moments later, SAROS continued,

"It appears that your eyes are just fine, Captain. The distortion is real. The same type of energy detected on the ocean floor also exists within the atmospheric distortion."

"Any ideas what it may be, SAROS?"

"There is too much insufficient data to give a reasonable speculation at this time, Captain. I will continue to look into it. As I receive new information, I will report it to you."

"Thank you, SAROS. How about that visit to Marathrax while I eat?"

"My pleasure, Captain! Setting course!"

"Engage!"

As the ship ascended, I gazed dreamily at the planet's surface. I thought of how tranquil my visit was, and promised myself that I would return.

Chapter 19
On Marathrax

While heading to Marathrax, the Lotania flew past its only satellite, which SAROS called Varyn. From our distance, no major detail could be made out, except to say that it was barren in appearance. As the ship continued, it began its descent to the surface of Marathrax. As it did so, I kept myself from looking out of any windows. I then made a request.

"SAROS, could you please relay to me the planetary conditions of this planet?"

"Yes, Captain! Marathrax has an oxygen-nitrogen atmosphere, with other trace gases similar to Earth. No spacesuit will be required when you exit the ship. Like the 3rd planet you recently visited, Marathrax has a slightly higher oxygen level. This may cause you slight dizziness until you adjust. As for the barometric pressure, it falls within acceptable limits for your body. The current temperature is 67 degrees Fahrenheit. The gravity is 87% of Earth's, or 8.524 m/s2. You will get used to the change rather quickly."

"Thank you, SAROS."

"You're welcome, Captain."

Still denying myself a glimpse of the surface, I took a few deep breaths to prepare myself for what remained on Marathrax. Gathering up the nerve, I opened the door and exited the ship. I beheld a city in ruins, but to my astonishment, there were no bodies to be found. It was as though nobody had ever lived here. It wasn't exactly what I had envisioned for my first contact.

I noticed that not far from the Lotania was another craft, very much like it in appearance. I then asked:

"SAROS, is that Richard Gaber's ship?"

"Yes, Captain! It is a twin to the Lotania in almost every respect."

"Is Richard inside the craft?"

"No, Captain! I do not detect the presence of Mr. Gaber anywhere in the area."

"I need to go and look for him, but first, I will check out his ship. Please hold down the fort, SAROS. See if you can find out what happened to everyone."

"Aye, Captain!"

Knowing that Rich wasn't in his craft, I took a moment to look at the city once again. As I already mentioned, it wasn't looking too healthy. The whole place was completely deserted, which confirmed SAROS' report about all life signs disappearing. I still couldn't wrap my head around it.

As for the design of the city, there were a number of buildings that were built in a circle around a central complex that stood taller than any of the other buildings. On top of the central structure was some form of time piece, which read 18:07:25. This device was still in operation despite how deserted and damaged everything was. The central complex was square, with a domed area at its center that had been completely damaged. Vegetation adorned the city, both within its boundaries and outside of them. Like the rest of the city, most of the plant life showed signs of trauma.

The buildings surrounding the central complex had geometrical shapes, each having taken some degree of damage. The Lotania landed outside the circle of buildings near Richard's ship. From the Lotania's position, I saw an impressive pile of rubble directly ahead of me. Very little of the building remained standing! From what I could make out of what was left, I'd say the building was once cubicle. Fragments of it now lay in the roadway. In the 7:30 position from the Lotania stood a cylindrical building. It took some damage but appeared structurally sound for the most part. Moving on to the 9 o'clock position, I saw a building in the shape of a pyramid. One side of this building took heavier damage than the other sides. At the 10:30 position was a building in the shape of a rhombus. This building, although not as tall as the central structure, most likely had the greatest area of all the buildings. It was actually quite massive.

Standing 180 degrees from the Lotania was a tall spiral-shaped building. This structure had taken a fair amount of external damage. On the roof of the building was something written in the Duniy language. *Oh, no, not again. Dirnoy, Donysh, Oh, forget it!*.......When translated, I found that it read Varyn Estates. A reference to Marathrax's satellite! Very interesting!....Dynersh, Deernog,.....I'm hopeless!

In the 1:30 position, I saw another pile of rubble in the roadway and into the vegetation beyond. As for the shape of the building, there wasn't anything left from which to make a guess. In the 3 o'clock position, I saw more rubble strewn about. From what little was left, I guessed that the building may have been rectangular. I could see the building had an outer and an inner wall, which would have left a sort of courtyard in the middle. This area of the city may have been hit the hardest based on the damages seen. Finally, in the 4:30 position stood a spherical building, which looked relatively undamaged. I needed to investigate these buildings in detail after I looked over Rich's ship.

CHAPTER 20
THE LOTANIA'S TWIN

Before turning away from the city, I tried to imagine what it was like without damage and bustling with people. I was dealing with multiple feelings within myself, all trying to get out at once. Trying not to be selfish, I couldn't help feeling robbed of a once-in-a-lifetime experience. I came so close. I was within an hour of meeting an extrasolar race of beings. Almost two billion people were gone just like that. I hoped to be able to get some answers to what happened here. After pondering for a moment, I looked over at Richard's ship. I then began making my way over to it.

Upon arrival, I cautiously entered, only to find it deserted like the rest of the city. Looking about, I could see that the ship was identical to the Lotania, except for a device hooked into the main navigational system. *"I'll be interested to know what it is when I catch up with Richard... that is, if I catch up with Richard",* I thought. No, I needed to stay positive. There's an answer to all this; I just had to find it.

In Rich's clothing cabinet, I found a few cool items that I didn't have. The most impressive were a laser pistol and rifle. I found a sonic grenade in the toolbox with protective earplugs, and a functioning timepiece similar to the one on top of the central complex. It seemed Mr. Gaber held out on me a bit when stocking my ship with equipment. The timepiece now read 18:23:38. I was not sure if it had been 16 minutes since seeing the readout on the rooftop, but it seemed reasonable.

Under the control console on the floor, I found another device I didn't know. It appeared to have been left on, still doing something. On the chair was a device similar to a communications device used at home. If I wasn't mistaken, the device had a message waiting to be seen and heard. Activating the unit, I heard explosions going off periodically. Rich was seen within the ship as he left the following message.

"Captain Travis, I do hope that you receive this message. It would mean that you successfully made your journey to Marathrax, for which I highly congratulate you. I regret, once again, not being able to greet you in person with the Dhirniy leaders as I originally planned, but unsuspected misfortune has struck this wonderful world. The race that inhabits the second planet in this solar system is attacking the Dhirniy city outside my ship at this very moment. I have learned that the contention is because of the third planet, and the Dhirniy's advanced technology."

"The third planet, as you may already know, has a surface of water except for a small segment of land. The Allorayans, as they are called on the second planet, want to claim the water planet for themselves due to the harsh and dry conditions of their own planet. They feel this will help them survive as their planet dries up over time due to its close proximity to the Sun. The Allorayans also fear the technology the Dhirniy have achieved. The Dhirniy have successfully constructed a device that can alter time, enabling them to travel through time. The Allorayans feel threatened by this and wish the technology for themselves, or to destroy it if they can't get it. Therefore, I suspect that in the minds of the Allorayans, their attack upon this city is for the purpose of continuing their survival. Although they have no need to carry out their fears, I can understand their position."

"I was in a large building in the city's center when the attack began. It is there that the time-altering device exists. It is under the domed section of the building. The device can only operate when five keys having geometrical shapes are input into the mechanism. The device must be active before the keys can be input. At the time of the attack, the device was active. When the attack began, I immediately removed the five keys and hid them safely away. I did this so the Allorayans could not steal the technology and use it as a weapon against the Dhirniy. I managed to avoid the Allorayan's attempts to stop me from hiding the keys. This was not an easy task, seeing that I had to go from place to place within the city to hide the keys. It took me over an hour to accomplish this action; seeing my movements had to be carefully made and timed.

When done, I made for my ship, hoping to escape to a safer area. Unfortunately, I could not do this before the attack became a bit more centralized in the vicinity of my vessel. I immediately raised the shields on my ship, but know that I have very little time left to leave this message. My shields are weakening, and soon, my ship will be defenseless."

"My friend, you must find the keys I have hidden away and keep them safe. Take the universal translator I've placed on the control console before you leave. With all five keys, the city can be restored again by going back in time to a point before the attack took place. This time, we can see to it that the attack never happens."

"The keys are scattered throughout the city. I wish I could be more specific as to their locations, but I must take precautions in the event this message ends up in the wrong hands. To add extra security, the keys can only be retrieved by solving a puzzle or two that I created some time ago as a safeguard to their protection. I have also left messages behind to aid you in your quest. Some of the messages will not be understood by the Allorayans, this should, therefore, be —(an explosion is heard that rocks the ship)—that's it, my shields are down, and I'm about to be boarded. Good luck, my friend."

I saw Rich rise from his seat and drop the device on the chair. Moments later, another voice spoke,

"Step away from the control panel, and come with me."

I couldn't see the face of the speaker, but I did see Rich get grabbed. Sounds of a struggle took place off-screen for about ten seconds, and then all went quiet. I watched a little longer to see if anything would follow, but there was nothing more.

Startled by what I had just seen and heard, I tried to gather my thoughts on how to proceed. I gathered the items Rich made mention of, and placed them with my gear. Rich told me that I needed to find five geometrically shaped keys that could restore the city somehow. This was a lot to take in. Going back in time to correct problems was only something I've ever heard about in movies.

Wasting no further time, I began putting a plan together and got underway immediately. I hoped to succeed. I didn't want to let Rich down, or anybody else for that matter.

Chapter 21
Murals on a Cylinder

Before leaving Rich's ship, I gathered the deciphering code and all of Rich's weapons. Once outside, I chose to start my investigations with the cylindrical building. I had no particular reason why. I guess I just started with the building closest to the craft. I would move clockwise from there.

Walking through the city was exciting, but a bit eerie as well. All I heard to break the silence was the crunch of the ground under my shoes, and a wind that whistled as it passed my ears. When I arrived at the cylindrical building, I cautiously entered. Looking around, I could see damages from the attack that recently took place. There were tipped-over statues, some that had broken on impact with the ground. I also saw other artifacts scattered about the floor. Many pieces of artwork displaying people and places adorned the walls. The building looked like it could have been a museum of some kind.

In the middle of the building, as though it was its support beam, I found a spiral staircase. All around the wall above the hanging artwork was a large mural of the solar system belonging to the star Uganthos. The mural displayed a giant orange star, seven planets, and the satellites belonging to each of the planets. I saw an asteroid field between the fifth and sixth planets. Planet 4, the home of the Dornie, had only one satellite. Planets 5, 6, and 7 also displayed satellites. Planet 5 showed three, planet 6 had seven, and planet 7 had two. Planets 1, 2, and 3 had no satellites.

Around planet 4's satellite was a halo, or ring. I did not see any such structure around the satellite when the Lotania flew past it on its way to Marathrax. I can only, therefore, believe that there is something of significance about the satellite SAROS called Varyn.

Ascending the stairs, I entered another room with quite an impressive display of artwork. Unfortunately, this floor took greater damage than the first floor. There was a good amount of rubble lying about that used to be machinery and statues. A large portion of a decorative window had been broken, scattering its fragments all over the ground.

This floor also displayed another remarkable mural. On this work of art was a dry and barren landscape showing a similar appearance to the surface of Earth's satellite. I surmised that I was looking at a likeness of Varyn. Of special interest, I beheld that the artist focused on a small outcrop of rock encrusted with minerals. Not far from the outcrop stood a cliff side of considerable height. I saw on the mural another small halo encircling an area of the outcrop's upper portion. I also noticed a small bright spot being displayed on the wall of the cliff. I took the two haloed areas from both murals to be related. The first showed Varyn as seen from Marathrax, and the other was a specific area of interest on Varyn.

To the left of the outcrop of rock, I saw an unpainted area of the mural. Within it, I found nine characters written in what I suspected was the Durney language. Checking the deciphering paper, I confirmed my suspicion. The characters were the numbers 529053118. I assumed this number sequence to be another coordinate, possibly to the area depicted on the mural of Varyn's surface. So that I would not forget, I wrote the numbers down. I then headed back to the first floor.

While descending the stairs, I caught sight of a smooth, gray stone tablet standing securely on the ground near the wall. It was positioned where I could not see it when I first entered the building. The stone stood about two feet in height, and on the tablet was the Dernish symbol for the letter B. Jotting this information down, I knew I had another mystery on my hands.

As I prepared to leave the building, I heard the sound of something hitting the floor upstairs. With all the damages the building recently took, it could simply have been something that had a delayed reaction fall. Nevertheless, I thought I'd take one last peek to make sure. Slowly ascending the stairs again, I looked around with greater detail. After a minute or so of moving about very stealthily, I heard another small disturbance. This time, it sounded like the shuffling of someone's movements. Moving quickly to where I heard the sound, I came across the very thing I'd hoped for: almost. I was staring into the face of another being, a being who was holding a weapon on me. He spoke and said:

"Who are you?"

"You speak English!" I responded.

"I said, who are you?"

"It must be the translator," I spoke aloud to myself.

"Ah! My, my name is Captain Lee Travis from a planet called Earth, which is about 370 light years from here. I was invited by your president to come here, and certainly didn't expect a welcome like this from the only person I've seen so far."

Then the stranger continued, "Are you responsible for the disappearance of everyone that used to live here?"

"I assure you that this is how things were when I arrived here." "You are the first person I've seen. Why is this city in ruins, and would you please lower your weapon? I mean you absolutely no harm."

The alien slowly lowered his weapon and continued.

"The city was attacked by the race of beings that live on the second planet. The planet is called Allorayus. I'm sure that many other cities like this one were also attacked."

"Why would they want to attack the people of this planet?" I asked.

"Because of the third planet, which is mostly water," responded the stranger. "Allorayus is drying up because of an occurrence that affected its axial tilt, and because of how close it is to the sun. They wish to claim the water world as their own; it seems their only means of survival."

I then responded, "Surely, there must be enough water on that planet for both worlds to share. There must be other reasons."

Apparently, I said something to cause the being to momentarily stare at the ground as if in thought. I then continued, "There is more to this, isn't there?"

The stranger responded, "You are very perceptive, Captain Travis."

The stranger then exhaled deeply and continued.

"There is a device on this world that can alter time, and the Allorayans are very uneasy about it. It is under the dome within the central complex. The building is very well protected so that the device can be kept safe. Nonetheless, I believe that the attack on the city knocked out the safeguards used to protect the device. If this is the case, I also believe that the device has been damaged. The damages could be responsible for the disappearance of the people of this world."

"How can the device be responsible?" I asked.

The stranger responded, "As I said earlier, the device alters time. I believe that during the attack on this world, the device was affected, and possibly shifted the population of this planet out of phase with normal time."

"Correct me if I'm wrong, said Captain Travis, but are you saying that the people are not really gone, but are simply occupying a different time frame than ours."

"Exactly, Captain! They can be out of phase by as little as a fraction of a second, and it would be enough to render them invisible to our sight."

"How is it that you escaped the time sweep?" I asked.

"I don't know, Captain," responded the stranger. "I was working in a closed metallic area in another building during the attack. It must have protected me somehow. I was lucky."

I paused for a moment, pondering all that I heard, and then I said, "Do you have a name?"

The alien chuckled at my question and said, “It’s Noks; please forgive me. With all the excitement, I…”

“No need to apologize; I understand.” I then continued:

“So, Noks, how do we proceed?”

Noks answered, “I figure if we can get the time machine to function, we might be able to bring everyone back.”

“Let’s go have a look at the machine,” I said.

“We can’t,” Noks responded. “We need special keys to open the door it is behind.”

“You don’t have the keys?” I asked.

“No,” Noks responded. “I do not work for the company within the central complex. But I do know that the key combination requires at least three keys that must be placed in their proper order just to enter the building. Trying to open the door without the keys would be a waste of time. I began looking for information that might help me locate the keys. That is when you entered the building and I found a hiding place, thinking that the enemy had returned. When I saw that you were not of this world, or from Allorayus, I watched you. I did not wish to shoot a possible friend.”

“I am grateful that you were interested in talking more than shooting,” I said. I then continued:

“I suppose that we should continue looking for information that can help us.”

Noks responded, “Perhaps if we split up, we will find more. We can meet up with each other a little later.”

“We can do that,” I said. “Good luck in your search. Oh, and Noks, I’m glad you were not caught in that time sweep.”

“Thank you, Captain; me too.”

After Noks left the room, I thought about the incredible conversation I just had. I was actually speaking with a being from another world. It was nice to have someone else to talk to outside of SAROS. Not that there was anything wrong with SAROS. It’s just, well, I hope you know what I mean. I find it difficult to put into proper words.

With the room all to myself now, I gave it a careful once-over. While searching, I came across a map of the sky that was inside a stone box. From the look of it, the map had already been discovered by Noks, who apparently seemed to have shown very little interest in it. The map highlighted nine constellations, and it was my job to find out why. Finding nothing more, I placed the map with my gear for possible future reference and left the cylindrical building.

CHAPTER 22
THE PYRAMID PATH OF PRIMES

Continuing in a clockwise direction, I head over toward the pyramid-shaped building. I couldn't help noticing how peaceful it was here, and not because of the absence of the people. The place just had a peaceful feeling about it.

Looking around, I also noticed that I had no idea where Noks was at the moment, not that it was important. I guess I just missed having company. Talking with Noks filled an emptiness I was feeling.

As I approached the building, I saw a great deal of damage on the ground on its far side. *Such a waste,* I thought, as I entered the pyramid.

Once inside, I saw a passage that was narrow that ascended a bit. As it continued, it widened until it opened up into a good-sized room, which I figured to be the pyramid's center chamber. Inside were instruments that could be used to measure such things as length, time, and mass. Upon the walls were many writings, which, when decoded, turned out to be mathematical formulae. Many of the instruments were damaged, if not destroyed. Only a few of the artifacts remained unscathed. A number of the instruments were unknown to me as to their function and purpose. Even so, I appreciated looking around at all the devices of mathematics and science. I hope to remember the unknown items and later learn about them. Perhaps, I can bring the knowledge back to Earth with me and advance Earth's science and technology understanding.

From the pyramid's central room, I was given a choice of three possible directions in which to continue. The directional choices were right, left, and straight, all of which continued to ascend. I started with the left path, which led to a small chamber that had more artifacts strewn about. Pictures of individuals hung along the walls, possibly accomplished scientists and mathematicians of the planet's history. Stretched across the ceiling was an impressive three-dimensional model of the Uganthian solar system. Again, the satellite Varyn was shown with a halo around it. I wondered when I'd be able to see what all the excitement was about on Varyn.

Lastly, I found lying on the floor, another stone tablet like the one I saw in the cylindrical building. This tablet was broken in two, but the symbol for the letter Y was not damaged. Finding nothing more within the room, I exited the left branch and headed over to the path on the right.

Within the room on the right, I saw that part of the wall had been blown away, exposing the chamber to the outdoors. This damage explained the excessive rubble seen on the ground on the far side of the building. The remaining undamaged part of the wall was beautified with yet another mural of exquisite workmanship. It showed a man who stood in space, pointing into the solar system at the planet Marathrax and its satellite, Varyn. Just like the other murals, this one also displayed a halo around Varyn.

On the floor, there were more instruments of learning. I also saw a life-size statue of a man that had toppled over. The face on the statue and the face of the man in the mural looked very much alike. At the base of the statue were five characters, which, when decoded, displayed the name of Varyn. Apparently, the satellite orbiting Marathrax was named after this individual, making him an important figure in Marathraxian history. I hoped to learn more about Varyn as my quest continued.

After leaving the room on the right branch, I headed over to the only path left that I hadn't investigated yet. Back at the three-way intersection, I began to ascend the path that went straight. The ascent to the room was the steepest of the three. When I entered the room, I had to squeeze past a couple of large stones that dislodged from the pyramid's roof. I was able to see that I had reached the pyramid's pinnacle. The top of the ceiling at its highest point was, in fact, the top point of the pyramid. I also saw a trap door in the ceiling, which acted as a skylight.

As for the contents of the room, there really wasn't much of anything. What I saw was a ladder lying on the floor and a metal box sitting on a table. *Here we go again,* I thought. Something told me that getting the box wasn't going to be as easy as it looked. I say this because the floor leading over to the box was comprised of tiles, 50 of them, to be precise. The tiles were numbered in the Dihering language, and arranged in a five-by-ten pattern. The tiles numbered one through five were closest to me. On both sides of the tiles was a drop to the bottom of the pyramid. At the end of the tiles was a small stone path, which had the width of three tiles. The path continued until it reached the area with the table and box.

Just before the tiles began, the floor was like a platform of flat rock from which to start across the tiles. On this area of rock was sketched a message written in the Dynerig language. It had been badly damaged by one of the stones that fell from the ceiling. Part of the message was chipped away, and was nowhere to be seen. I could only assume it had fallen into the abyss.

Translating what I could see of the message, I read,

"The Path...........me."

Apparently, my clue on how to get across the tiles was not going to help me. I could just try to make my way across, but that would be an act of recklessness. I had no desire to end up at the bottom of a pit. To check out my idea of recklessness, I picked up a descent-sized rock that broke off one of the fallen roof stones, and cast it onto the path of tiles. At first, all seemed solid as the stone moved along the tiles. When the stone came to rest, two tiles, numbers 32 and 33, collapsed under its weight. The tiles should have easily supported the weight of the stone.

No! Something wasn't right here, and there was no way I was going to try my luck without having a plan on how to get across. The box would have to wait a little while longer. I didn't figure it was going anywhere.

Out of curiosity, I retrieved the ladder and rested it against the wall so that I could get to the trapdoor. I wanted a look outside from the top of the pyramid. Climbing the ladder, I opened the door in the ceiling. Looking outside, the city was still as empty and silent as it was before. After a few moments, I was about to close the trapdoor, when I saw that the pyramid had a set of stairs built into it leading to the ground. From the outside, I could not see the stairs because it was on the opposite side of the building from where I entered. Taking advantage of this good piece of luck, I carefully exited the pyramid building via the roof, and descended the stairs to the ground below. Continuing with my clockwise itinerary, I headed over to the rhombus building.

CHAPTER 23
THE ROTATING RHOMBUS ROOM

As I approached the rhombus structure, I looked back at the pyramid building and saw the area that had been blown away. Not far from the building was a metal statue that may have caused the damage to the pyramid. If the statue was launched by an explosion, the force of its impact on the wall could have done it. Looking at the statue more closely, it could have been another likeness of Varyn.

When I arrived at the entrance to the rhombus building, the door was hanging partly off of its hinges. Carefully pushing it open, I walked into a foyer that led to another door, which was rather ornate and in good repair. Stepping through the second door, I entered into a large octagonal room, which was probably the center of the rhombus building. Within the room were eight doors, each with a drawing on the door. I also came across another stone tablet, this time with a Derney symbol representing the letter 'K.' The drawings on the doors were quite interesting.

From my point of entrance, which I will call the six o'clock position, I saw the following drawings. In the seven-thirty position, the door had formulae drawn all over it. This is a room that people who fear math would gladly steer clear of. On the nine o'clock door, I saw a rhombus-shaped building with a path around it. The path also had an extension jutting off of its back right corner. I can't say where the extension led; the drawing did not show any details. The path must represent the walkway that goes around the building. I only say this because of the walkway I saw encircling the building before I came in.

In the ten-thirty position, I saw a door loaded with constellations seen in the night sky. The patterns looked much like those I found on the map from the cylindrical building. The door in the twelve o'clock position had a spiral-shaped organ-like musical instrument displayed on it. Such an instrument would be remarkable, to say the least. *If this drawing represents an actual instrument,* I thought, *I can't wait to see it.* Continuing on, the door in the one-thirty position showed a picture of Varyn. It appeared that no matter where Varyn was displayed, it was done as if being reverenced.

The three o'clock door showed the same thing the nine o'clock door displayed. The door in the four-thirty position had four colors within a square. The square was divided into four equal-sized smaller squares. The colors were violet, positioned in the upper left square, red in the upper right square, blue in the lower left square, and green in the lower right square. In the middle of the four smaller squares was a small white circle.

Finally, on the door at the six o'clock position where I entered the room, I saw a picture of the domed central building located in the city's center. Seeing that investigating all these rooms was going to take a little while, I got started right away.

Beginning with the door in the seven-thirty position, I found myself in a room that seemed dedicated to the pyramid building. I guess that was why the door had formulae all over it. The pyramid building had artifacts and instruments of mathematics and science. In the room, I saw a very large machine that took on the appearance of a computer, only much more sophisticated in its design. The walls of the pyramid room were like large viewing screens with doors located at their lower right corners. On the computer-like machine, there were four buttons in the form of a triangle, each one a different color. One button was blue, located in the top position; one was red in the lower left position, and one was green in the lower right position. The last button was white, located in the middle of the triangular figure.

The machine appeared to be inactive. Pressing a few of the buttons with no results confirmed my suspicion. Unsure of how to proceed, I investigated the doors on each wall. There weren't any doorknobs on the doors. Instead, each had a lit button off to the right, which I opened by touching the buttons. Behind each door was a smaller version of the large machine. The machines were identical to each other, except for a single colored button that was on the machine. The colors on the smaller machines matched the colors on the main machine.

I stood in the room with the machine that had the blue button. I assumed I had to press the button, and therefore, I did so. Again, it wasn't one of my more brilliant moves. Anything could have happened when I pressed that button. But what did happen, though, was that the button lit up. Seeing that I didn't die by pressing the button, I went to the other two machines and pressed their colored buttons, also. I ended up feeling that I foolishly played a game of Russian roulette. I still could have triggered something bad by pressing the other buttons. Thankfully though, the results were the same. Pressing the buttons only illuminated the lights. Being somewhat disappointed that nothing happened, I took a step back to think of what to do next. Perhaps the buttons needed to be pressed in a certain order.

With the three buttons on the smaller machines now pressed, I noticed that the main machine's lights activated. I guess something did happen after all. I, therefore, continued with my reckless behavior and pressed the lights on the main machine. The results of my actions yielded absolutely NOTHING. I pressed them a second time with the same results. *What was I supposed to do?* I couldn't let these puzzles stump me all over again. You'd think I'd be used to them by now.

Thinking things through a bit, I didn't come up with much. It wasn't until I recalled my own recently spoken words that I found a hopeful way to proceed. I was remembering my numbskull actions of randomly pressing buttons when I remembered I spoke of order. It seemed that the machine didn't care how I initially pressed the buttons, just how I pressed them the second time. I pressed red, green, and blue the two times I got no results, but that's not how I pressed them initially. The correct order should be blue, green, and red. With something to hope for, I pressed blue, green, and red on the main machine, and nothing happened. *Oh, come on! Now what?*

Staring at the machine, I was certain that I input the correct order for the lights. Continuing to gaze at the lights, my eyes went out of focus. *I must be tired again.* When I refocused, my eyes rested on the white light in the center of the triangular figure. Of course! I needed to press the white light as well. It wasn't there simply for decoration. That was the ticket!

Completing the sequence made one of the view screens come on. A message written in the Dynerg language appeared. After translation, the message read:

"The path is prime."

OH HOH! Would you look at that, I said aloud.

What an unexpected but beautiful sight. It was the full clue from the pyramid puzzle with the tiles. Now, I could hopefully figure out what to do. I was about to leave the room and head back to the pyramid building when the other two view screens began to display images. Mesmerized, I watched a recorded playback of the attack on the city. It was disturbing to watch, to say the least. I witnessed the crafts from the second planet come in and open fire on the city. I wondered earlier what the city looked like while bustling with people. I received an answer to my question; up until all the people just suddenly vanished, that is. It was wild! I mean, one moment, they were all there; the next, they were gone. Just like that! They were all gone.

Just before everyone vanished, I remember seeing the central complex building get hit by a volley of torpedoes. The hit was directly on the dome, smashing the glass to pieces. Rich told me in his message that the time-altering machine was located under the dome. It must have been damaged during the attack, causing the temporal disturbance to take place.

I suppose Noks was right. The people were not dead; they were just out of phase with normal time. That was mind-blowing to think about. That meant they were all still here, just slightly out of phase with time. *Ooo! I wondered how many of them I had walked through so far. That was pretty creepy!* Nonetheless, it brightened my spirits to know that the inhabitance were not dead, and that they could be brought back. I needed to hurry and get the keys if the city was ever to be restored.

An interesting part of the video was when the images were displayed from many locations, and from different angles of the locations. I say this because some of the images were of well-protected areas within the structures. When I spoke with Noks, he told me that he wasn't affected by the time sweep because he was in a well-protected location. All of a sudden, Nok's story didn't add up. According to the video, all protected areas were affected like any other area. I had to keep my eye on Noks. There was something that he was hiding.

Before the disappearance of the people, the video showed Rich's ship being entered, and then Rich being forcibly removed. He and another person were made to enter one of the enemy craft, which then lifted off the surface. Three other ships followed closely behind the lead craft. Those four craft must have been the ones I spotted from space. One of them turned to engage me. I hope Rich wasn't on that particular craft, seeing I destroyed it. That would really stink if all I did was come all this way just to flame out my friend. I didn't really believe the craft I fought had Rich on it when I stopped to think about it. Rich's ship left first; the craft that fought me was not the lead ship.

After the four craft left the surface, everything was silent except for the sound of the wind as it blew over a deserted and fragmented city. I stood there mesmerized, watching the screen, but nothing more happened. When the images stopped playing, I turned and slowly made my way back to the pyramid building. It wasn't easy taking in those images. Seeing acts of meaningless violence never are.

On a lighter note, I was able to see that the three buildings that were reduced to rubble were cubicle, rectangular with a courtyard in the center, and octagonal. I got two out of the three correct. There wasn't much left of the octagonal building to even make a guess.

CHAPTER 24
THE CONQUERED PYRAMID

Coming back into the main octagonal room, I saw seven other doors that needed to be investigated. But before doing so, I left the rhombus building and headed back over to the pyramid building to attempt the tile puzzle. Walking back offered tranquility to the mind and devastation to the eyes. It was hard to know how to feel.

Once back at the pyramid building, I entered it by climbing the outer staircase that led to the trapdoor on the roof. Thankfully, the door didn't lock itself when I last closed it. When back under the pyramid's pinnacle, I felt I was certain what needed to be done to get safely across the tiles. "The path is prime" must mean that I needed to step on the prime numbered tiles in the correct order from one to fifty. I saw what happened when a mistake was made. I, therefore, figured out all the prime numbers between one and fifty before going anywhere.

The tiles were big enough to support a person's foot. As long as I wasn't clumsy, I shouldn't end up breaking any more tiles. Hoping I didn't miss any, I wrote down the numbers 2, 3, 5, 7, 11, 13, 17, 19, 23, 29, 31, 37, 41, 43, and 47. It was now the moment of truth. I stepped onto the number 2 tile, and it supported me. It seemed silly to step on tile 3 when I could easily reach tile 17 in the next row, but I was going to be patient and follow the rules. Therefore, I stepped on tile number 3 next, and it supported me. Although it seemed to take an unnecessary amount of time, I obediently stepped on each tile in the correct order. After stepping on tile 47, I was able to safely step onto the path that led to the box and table.

Walking carefully, I made my way across the skinny path. Seeing there were no extra puzzles needed to open the box, I freely did so. Looking inside, I saw a beautiful, four-inch ornate key with the shape of a pyramid on one of its ends. It was finally coming together. I had one down, with four to go. Back to the rhombus building!

Chapter 25
The Answers to the Puzzles

Once back in the octagonal room, I continued moving in a clockwise direction. The next door was the one positioned at nine o'clock. Opening the door, I gave myself quite a startle. Expecting to walk out onto the walkway and over look the city almost cost me greatly. This was due to the fact that there was no walkway to walk onto, only a fifteen to twenty-foot drop to the ground. Apparently, the walkway was blown away during the attack on the city. If I had taken the time to examine the outside of the whole building before entering, I would have known.

Backing away from the opening, I closed the door. Taking a moment to gather myself, I then continued to the door in the ten-thirty position. Inside, I saw that the room was completely spherical. There was nothing inside of it at all, just a white wall with small black dots. Not long after I entered, the whole room began to rotate. It was a bit unnerving and disorienting. The room rotated about a quarter of a circle. When the room stopped, it exposed an entrance to an outer room. I approached the external room to get a better look at it, but before I was able to go inside, the room began to rotate again. When the room stopped again, I jumped back in a hurry. I found myself staring out of a damaged area of the wall, an extension of the area where the walkway used to be. That was the second time I found myself practically hovering over empty space. I got the feeling this building didn't like me very much.

Thankfully, the room began to rotate again, this time stopping where it should have in the first place. Wasting no time, and again, not thinking very clearly, I stepped into the outer room. I say that I wasn't thinking very clearly because I could have ended up trapped in the room. I'm very happy to say, though, that fortune favored the foolish. The room remained stationary while I was in the outer room.

On the wall, I saw a set of three constellations. The wall also had a number of deep cracks due to its proximity to the damaged wall off to the left. Wishing to be in the room no longer than was necessary, I jotted down the star patterns that were displayed. The patterns were of a sphere having sixteen stars, a figure of a man having seventeen stars, and a rectangle having eight stars. Having completed my task, I quickly exited the outer chamber. About five seconds later, the room rotated another 90 degrees. It stopped, revealing yet another room like the first two. Stepping inside, I noticed three more constellations displayed on the wall. This time, the patterns were a spiral constructed of eighteen stars, a pyramid with six stars, and a cube with eight stars. It seemed the patterns were related to the buildings in the city.

While drawing the constellations down, I heard the alarming sound of masonry breaking apart. I later saw that the room with the second set of star patterns crumbled away. The deep cracks in the wall gave way. The damages were too extensive for the room to hold itself together. I'm just glad I wasn't in there when it happened. This building was freaking me out. I swear it had it in for me.

Leaving the outer room, I couldn't wait to leave the entire sphere, but I had to get the first room I missed. Of course, that meant rotating past the damages again to get to the main entrance, but at least this time, I was ready for it. On schedule, the room rotated to show the main door to the sphere, and then the room I missed. Entering the room this time, I drew down the three final constellations. The patterns were of a rhombus constructed of twelve stars, an octagon of sixteen stars, and a cylinder of fourteen stars. When finished, I darted out of the room to get as far from the damaged wall as I could. The room rotated as expected; only this time, it did not stop until it reached the location of the third room. *Figures! I wasn't standing near the edge, so why bother to stop? I really didn't like this building.* When the room rotated to the main door, I did not hesitate to leave. Good riddance!

Feeling very glad to be back in the octagonal area, I headed for the door in the twelve o'clock position. Looking at the drawing again, I hoped that I would have the opportunity to see an actual instrument like the one displayed. With a small smile on my face, I entered the room. This chamber was related to the spiral building within the city. The entrance brought me into a corridor decorated with pictures and drawings that depicted music. The instrument was shown again in greater detail. Also, the spiral building was roughly introduced in the drawings, showing how to find the instrument. *Excellent! It does exist,* I thought. The instructions on how to find the instrument were rather confusing. Its location was shown to be the top floor of the spiral building. The drawing represented its location with dotted lines, instead of solid ones. After studying the diagrams and pictures, I continued.

At the end of the corridor was an apparent dead end with three buttons on the wall. Beneath each button was a Dourny character, which translated to the numbers 1, 2, and 3. I didn't have to look up the characters this time; I'm in the process of memorizing the letters and numbers. Now, if only I could memorize the races name. One of these centuries, I'll get it.

On the right side of the corridor, not far from the dead end, was another button. Pressing it granted me an earlier request I made. A recording began with a holographic representation of the individual named Varyn. A spokesman said the following:

"This city is happy to bear the name of our founder, Varyn Keb. He is a hero to be honored for his bravery and scientific skills. Varyn Keb predicted that the Sun to our former world would become a supernova much quicker than previously thought. He quickly alerted the Dhirniy people of the impending catastrophe. Although the Sun was known that it would soon supernova, it was believed that it would not do so for many millennia. Due to the scientific findings of Varyn Keb, his studies showed that vast instabilities existed within the core of the Sun, which would bring about the Sun's collapse within our generation. Heading the project to bring the Dhirniy people to a new home, Varyn Keb located this world that would sustain life. Varyn then led the Dhirniy people to safety. The people of Marathrax will always remember Varyn Keb, and how he cared unselfishly for the well-being and continuation of the Dhirniy people."

"Wow! A true hero indeed," I thought. *"No wonder they honor him so deeply, even so far as to name a satellite after him. He literally saved the Dhirniy race from eventual destruction." "Hey! What do you know!" "I actually said the name correctly again."*

After a few more thoughts about what I just learned, I turned to the three buttons on the end wall. It seemed I had a choice to make, unsure of what the results would be to either choice. I didn't worry too much, seeing that I didn't believe the Dhirniy people would bring harm upon me. Then again, the whole blasted adventure so far was loaded with the possibilities of impending death. Perhaps I should concern myself a little. *"Hey, that's two in a row. There may be hope for me after all."*

Regardless, I was expected to make a choice, or I could turn around and leave the way I came in. Not comfortable with making blind choices, I actually did make my way back to the 12 o'clock positioned door. I felt a little like a coward as my hand reached for the button on the side of the door. To my surprise, pressing the button did absolutely nothing. Trying again a second time yielded the same results. It would seem that in order to go back, I needed to proceed forward. *How nice!*

Heading back again, I looked at the three buttons and chose button number three. Oddly, door number one to my left sunk into the floor, revealing a black void. Peering into the blackness, I couldn't see a thing, no matter how close I got to the door. Even shining my flashlight into the blackness revealed nothing. Feeling very uncomfortable about door number one, which shouldn't have opened in the first place, I pressed button three again, but the door did not close. Hoping that another door would open, I pressed button number one. This time, door three to my right opened. *"This place has serious wiring problems,"* I thought.

Inside room three, I was able to see a dimly lit path. At least I was able to see something this time. Checking out button number two actually opened door number two. *Amazing! One of the doors worked properly.* Room two gave the best-looking option. I say this because the way before me was well-lit and much more inviting. A lot of times, though, the nicest-looking way ends up being the deadliest way. Nonetheless, I chose to take my chances with door number two.

Once inside, it seemed like a rather straightforward path. It offered no turns, or anything else for that matter. I noticed a transparent tube big enough for a person to fit in, running over my head as I walked. I wished I could be inside of it so I could see where I needed to go. It would have made things a lot easier.

After walking for about three minutes, I came to another apparent dead end with a set of three buttons. Seeing that I had good luck with door number two the last time, I foolishly chose it again. When the door sunk into the floor, I saw another path that was dimly lit. Wishing to find the brighter path again, I chose another door. To my disappointment and shock, neither of the other two buttons operated this time. Unhappy with my situation, I entered onto the dim path. My flashlight was needed in a few of the darkest areas of the path. I noticed that the tunnel tube was not over me. I guess I picked the wrong door. I just hoped that the path would lead somewhere beneficial. Eventually, after a number of turns and real dead ends, I came to the fake dead end with the three buttons.

To be merciful, I will spare you the repetitive and very dull details of my annoying walk through that insane labyrinth. I will simply say that I walked for hours, trying to use the location of the overhead tunnel as a guide. At one point, I ended up on a path that nearly killed me when the ground gave way, leaving me dangling over a pretty deep hole. Barely escaping, I wasted no more time on that path and returned to the intersection. Figuring I was stuck, seeing the other two buttons didn't function after my initial choice was made, I sat down for a few minutes to think of what to do. I decided to head back to the intersection before this one.

Before heading off, I hit button number three in frustration. To my surprise, the door to my right opened. It looked like the wiring problem was fixed. Happily, it was the bright path which eventually brought me to another fake dead end.

Blah! Blah! Blah! As I said before, I will spare you the coma-inducing details. Just know that I eventually found where I needed to be. I ended up in a room with, yes, you guessed it, a metal box. Inside was a.........care to guess again? If you said key, you are absolutely WRONG. Instead, the box contained a message written in the Dharnoy language, which read:

"OH! I'm going to cry." No, that's not what it read. The message said:

"Play the Founder's song."

I had no idea what it meant, but I hoped it would deal with that cool instrument. In my excitement to find the metal box, as though I hadn't found more than fifteen of them already, I failed to see that there was an entrance into the tunnel tube within the room. It was a bit late, but as the old saying goes, better late than never.

Entering the tube, I got an impressive view of the maze beneath me. I can't tell you how overjoyed I was that I didn't have to find my way back by going through it again. The tunnel eventually ended at a door that led outside onto the walkway. Of course! It was the walkway piece that branched off the main walkway encircling the rhombus building. If I had just walked out of the three o'clock positioned door instead of being so blasted systematical, I could have possibly skipped my whole maze experience altogether. To make myself feel better, I just kept telling myself that the three o'clock door would have probably been locked from the outside. To this moment, I still don't know if it was or not. I never went back to check.

Well, on to the next room in the one-thirty position. Pressing the button on the side of the door, it opened to reveal a cubically shaped room. Judging from my own height, I estimated that the chamber was about eight cubic feet. Other than the common sight of rubble strewn about, the room had very little in it. What it did have, though, was worth making mention of. This was due to the two beautifully crafted murals on the right and left walls. The left mural displayed Varyn standing outside, holding up a piece of parchment in the direction of the sun. The right mural showed Varyn looking up into a starry sky at the satellite named after him. They were remarkably done with extreme detail.

Other than the murals, there was just a piece of paper lying on the floor near the far right corner. I figured that I was expected to retrieve the paper, but the apparent easiness of the task was unsettling. With my luck, the floor was booby-trapped with mines or had areas that would collapse when stepped on. No! I wasn't going to just waltz out there without further understanding.

Looking around for anything odd or out of place, I couldn't find a thing. There wasn't anything in the room to find. Even the floor was level with no discernible deformities. The only oddities were the small holes in the murals. The most obvious were the ones situated in the Sun on the left mural, and the others within the satellite on the right mural. In total, I counted seven rather inconspicuous holes scattered about the room, three of which were in the ceiling.

Normally, these abnormalities would have gone completely unnoticed. But the apparent easiness of retrieving the paper was not to be ignored. It had 'trap' written all over it. *But what was the trap?* I wondered. I saw nothing to raise my red flag. Shifting my position around a little, being careful not to enter the main floor, my eye caught sight of a bright point of light. It came from the hole in the Sun on the left mural. Initially, I just accepted that it was the light from Uganthos shining through the window. The only problem, was that Uganthos wasn't on that side of the building. Looking very carefully, I tried to see if any of the other holes displayed any light source. Although hard to see, when I stood in just the right place while squatting down a little, I could detect a bright pinpoint of light coming from the Varyn satellite hole.

At this point, I assumed that all of the holes must have a light source behind them, but why? Walking back toward the door, I stopped and turned. Looking back, I saw dust particles floating within a beam of light that started from the Sun's hole. I could not see the full length of the beam, but I'm sure it was there. Again, only by positioning myself just right was I able to see dust particles in a beam coming from the Varyn hole. I said earlier that I couldn't see anything to raise my red flag, but just because you can't see something doesn't mean it isn't there.

The trap was invisible. Only the fact that there were dust particles floating in the air alerted me to the beams. I assumed that touching any of the beams would set something off. I didn't wish to find out what that something would be. The Sun and Varyn holes were about six feet off the ground and angled down about 40 degrees. Each mural also had another hole in it that went horizontally. One of the holes was four feet off the ground, while the other was two feet from the ground. The last three holes in the ceiling were spaced two feet apart from each other and shot vertically down. All of a sudden, the task of getting the paper wasn't so easy.

Drawing a rough sketch of how the beams were arranged, I planned my path across the room. The beams coming down from the ceiling were like bars on a cell door. A two-foot separation is tough to squeeze through. The beams shooting horizontally across the room meant that I must stay low. If not, I would need to step over the two-foot beam while squeezing through the ceiling beams. The beams coming down at angles meant that I must stay away from the area where they converged or any area close to it. No! This was not going to be easy at all.

According to my sketch, the best areas for me to attempt getting to the paper would be to the far right or far left of the room. I would have to move in between the first and second or second and third ceiling beams. I needed to squeeze through the two-foot space while staying below the four-foot beam. I also needed to avoid the two beams that descended on an angle by moving to the right or left of the intersection point. Lastly, I had to avoid touching the remaining floor beam that was shooting horizontally. *"This ought to be fun"*, I thought.

It would be too hard to describe every move I made to avoid beams I couldn't even see. I will say, though, that I felt like a contortionist, and……that I eventually made it over to the paper. Sweating like a bank robber in church, I picked up the paper to find THAT IT WAS BLANK. *"ARE YOU KIDDING ME! No! No! This was a joke of some kind."*

Totally in a state of disbelief and outrage, I threw the paper on the floor and mindlessly walked straight across the room to the door. I didn't even care about the beams at this point.

After all, what's the worst that can happen? Well, remind me NEVER to ask a dumb question like that again.

I didn't get three steps before the main door closed and locked. A few seconds later, the room began to fill with a pukey green-looking gas of some type. Running to the door, I checked to see if it would open; no luck! I tried kicking the door open, and I got the same result! I rammed the door with my body and received a bruised and very sore shoulder. In a growing state of panic, I tried to think of what to do. My lungs were beginning to fill with the unknown gas. Whatever I was going to do, I had to do it quickly. If only I had something to blow up the door. *WAIT! I did have something. While on Triton, I grabbed the remaining C-4 charges.*

Quickly! I placed one of the charges next to the door. I attempted setting the detonator to the correct frequency to match the charge as I began to lose consciousness. I prepared to hopefully blow open the door as I staggered away from it, collapsing to the ground.

BOOM!

The force of the explosion blew a sizable hole in the door, which I dragged myself through to escape the room. Choking on the gas in my lungs, I tried to breathe in the good air of the octagonal room. I saw the nasty-looking gas leaking out of the door. I sluggishly moved to a safer distance so that I wouldn't breathe in any more of it. When it appeared that the gas had dissipated enough that it was safe to continue, I slowly rose to my feet and staggered over to the door in the three o'clock position, which led outside to the walkway. Opening the door, I deeply inhaled a number of times, coughing out any remaining gas.

After the dizziness passed, I began making my way back inside. Still upset about the blank paper, I abruptly stopped. Of course! The mural, I said to myself! Quickly but cautiously, I approached the cubicle room again. I made sure it was safe to enter; then, I returned to the mural of Varyn holding up a piece of paper in the direction of the Sun. Varyn was showing me what I needed to do. Carefully avoiding the beams and retrieving the paper again, I took it with me outside the Rhombus building. Aiming it in the direction of Uganthos, I revealed an invisible message that, when translated, read the following:

"May all worlds unify as their vaults of friendship are exchanged."

It was a nice message, but I didn't know what it meant. I noticed there were letters underlined within the message, but I didn't know what they meant either. I wrote the underlined letters down separately, hoping they would make better sense, but they didn't. I should say, at first, they didn't. Playing around with the letters for about thirty to forty minutes, I arranged them and found another message, which read:

"Go to Varyn."

Wow! I hoped all the hype about going to Varyn was going to be worth it. I would have loved to go right then, but my task was not yet done. Storing the message in my backpack, I returned to the Rhombus building to finish my investigations.

Finding the door in the three o'clock position still open, I exited onto the walkway one more time. It was very pleasant outside, and I appreciated the few moments of peace. Already aware of the door leading to the tunnel tube and still dealing with leftover disappointment, I went to the walkway's dead end overlooking the city. It was a nice moment to relax and enjoy the peacefulness of a day that was nearing its close. Unfortunately, the moment didn't last nearly long enough. I still had a room to investigate.

As I turned to re-enter the building, I caught sight of Noks searching through the rubble of what was once the octagonal building. Still feeling unsettled about our conversation, especially after what I learned in the pyramid room, I watched him to see what he was up to for a few minutes. Becoming bored, I re-entered the building to investigate the room behind the four-thirty positioned door.

Ah, yes, the door with the colors on it. After entering the chamber, I saw that it was rectangular in shape and quite spacious. Each wall was a different color. The colors matched those that were on the door to the room. The ceiling was white to stay in harmony with the color scheme. It was a rather humid room due to the decent-sized swimming pool in the middle of the floor. What would I find next: a bowling alley?

The pool was about 14x11 feet, with a depth of 10 feet. The water looked rather inviting, especially after sweating so much in the other room. It just so happened that I would be going in for a swim. I say this because, I saw on the bottom of the pool, a tube being held down by two metal rods. Naturally, I needed to go in and get it. Even though I just wanted to jump in, I knew not to be that foolish. I tried to test the water for traps by sticking a pole into it. The pole didn't get eaten by acid, and it didn't freeze or boil. It seemed like a normal swimming pool, and that's what bothered me the most. Again, the task seemed too straightforward. I didn't have a good feeling about this at all. Being very cautious, I entered the water. After all, the tube wasn't going to come to me on its own.

OH YEAH! The water felt wonderful! For a moment, I remembered my visit to the water world and how relaxing it was. I began to take a couple of leisurely laps to wash my body and mind. I didn't realize how much I needed to do this. Feeling much better, I swam over to the area of the pool where the tube was being held down. Preparing to retrieve the tube, I took a deep breath and submerged myself. Not being a bad swimmer, I was able to reach the tube in about five seconds. Grabbing hold of the cylinder, I could not slide it out from under the metal rods. Having no choice, I lifted up on it, and the metal rods moved up as well. *Uh oh! What just happened?* Hearing something behind me, I turned to see massive air bubbles rising to the surface. This was due to a hole opening up on the floor of the pool. The hole grew steadily larger every moment. It was time to move.

With the tube in hand, I swam to the surface the best I could, but the opening was acting like a drain. I felt myself being pulled toward the center of the pool as I kicked toward the surface. Reaching the top, I watched as the hole grew in size. I found myself being pulled toward the pool's center. A whirlpool was forming now, and I knew I had very little time to grab hold of the side of the pool. Fighting with all my strength, which was waning very quickly, I took hold of the pool's edge. I almost lost my grip but managed to hold on. I held on very tightly as the water got pulled down a pipe that led to who knows where. I can assure you that I had no desire to find out. Soon, my body was out of the water as I hung on the side of the pool like a piece of wash on a clothesline. I watched as the water spiraled furiously into the hole, which was now about 2x3 feet. Pulling myself out of the pool, I rolled onto my back, exhausted. Amazingly, I was still holding the tube. A few minutes later, I lifted myself up and watched as the last of the water disappeared down the drain. About five seconds after the water was gone, the hole in the pool began to close again.

I watched the hole close, giving myself a few extra moments to rest. When all was quiet, I looked down at the tube in my hand and opened it. Inside was a water-soaked piece of parchment. The water leaked into the tube, ruining the image on the paper. All I could see was a fragmented outline of something, and a number of smudged, colorful geometric shapes. I risked my life again, and for what? How was I supposed to get anything from this water-damaged piece of waste? Granted, I had no idea that the tube was damaged. How could I know? Well, there was no need to get myself more worked up. I'll just have to chalk this terrifying experience up to having been a bad job, and move on.

Getting to my feet, I gladly left the room and entered back into the main octagonal chamber. Having investigated all the rooms, I finally, and happily, exited the Rhombus building. It was nice to get outside again. The Sun had gone down during my swim, leaving a peaceful-looking sunset of reds and purples. Giant orange stars leave a different type of sunset, but it was absolutely beautiful nonetheless. Being actually able to experience day from night for the first time in a while, I decided to turn in for some much-needed rest aboard the ship. In the morning, I would continue with the spiral building.

Chapter 26
Varyn Estates

The night passed very restfully! It was as if the entire evening were no longer than a few minutes. Before I knew it, I found myself awakened by a globe of fiery brilliance. I was able to hear the wind blowing outside the ship. It must have been a pretty strong wind to be heard within the craft. Despite how short the night seemed to be, I felt quite rested. I grabbed a quick bite and gathered my gear for the upcoming day. I would be starting where I left off last night.

When I exited the craft, the wind nearly made me lose my balance. The sky looked turbulent, showing areas of dark clouds that moved rapidly across the sky. For all I knew, a storm could be coming. I, therefore, made my way over to the spiral building named Varyn Estates as quickly as I could. I learned that the shortest distance between two points is a straight line. I, therefore, cut across the circular shaped city instead of walking around its perimeter.

As I mentioned before, the spiral building took a pretty good amount of external damage. It was a different matter inside the building. The internal appearance was quite impressive. There wasn't nearly the amount of damage I expected, based on the building's outer appearance. I felt like I walked into an upper-class hotel. To my left was an open area with tables and chairs. The room was richly decorated with art of various types. Sofas were against the walls, and holographic imaging devices sat atop each table. I know the devices were holographic imagers because I played around with one of them. Very cool! The scientists of Earth really need to learn how to build these things.

In the center of the room, which I'll call a lobby, I saw another stone tablet. On this one was the Dewney symbol for the letter R. On the right side of the room was the main desk with another holographic imager, only this one was much larger. I was able to see a holo-image of the rooms in fantastic detail. Those things were so cool! I'll have to get myself one of them!

Behind the main desk were cubby holes for storing items. In one of them, I found a piece of parchment. On the paper was a similar picture like the mural on the wall of the cylindrical building's second floor. It showed the landing site for anyone visiting Varyn. A sizable outcrop of rock was encrusted with minerals, and the rock face from the mural was also seen. I noticed a few things in the picture that were not shown on the mural. One was the giant orange sun rising off to the left of the outcrop. Another was a beam of light originating from the spot on the outcrop that was depicted with a halo. The beam of light was seen going over to the rock face, hitting it on the spot that was shown as a bright spot on the building mural.

I was able to deduce that the beam of light was caused by the light of the rising sun hitting the minerals on the outcrop of rock. The information on the paper stated that the beam occurs every day at the same time and lasts for thirty seconds. The beam energizes the depicted spot on the rock face, causing it to glow for a couple of minutes after the beam stops. Touching this spot while it still glows opens up new doors for you to explore. It went on to mention that the beam starts precisely at 81:99:70 rotations and last until 82:00:00 rotations. Lastly, it instructed all visitors to arrive no later than 81:75:00 rotations.

When I first arrived on Marathrax, the time was 18:07:25 rotations, as displayed on top of the domed central complex. Pulling out the timepiece I picked up from Rich's ship, I saw that the time was now 29:10:64 rotations. Oh yes, I also noticed one other thing on the paper. I saw an arrow pointing upward into the sky off to the right side. I would need to see what item of interest is visible in the sky when I get there. To make sure I did get there, I needed to keep an eye on the time so as not to miss my Varyn excursion.

Now, continuing with the building's description, there was a large spiral staircase in the center of the floor that moved like an escalator. The stairs went around a massive column measuring about five feet in diameter. When the stairs reached the top, they moved a little ways past the column and began to descend vertically until they reached the floor. Then they moved back to the column to make their way back up again. It was a fascinating design to say the least.

At the column's base, there was a door with no apparent way to open it. Looking up, I could tell that the stairs went up quite a ways. Stepping onto the staircase, I went to the first floor with rooms. There were seven rooms on the floor, each with a Derdish character on its front door. I recognized the characters to be the first seven letters of the alphabet: A through G. I assumed the remaining floors would have rooms H through Z, provided there were twenty-six rooms. Except for the fact that the hallway was circular, it looked like your typical luxury hotel; not that I've ever been in one.

I walked over to room A to check it out. Opening the door, I entered to see a levitating bed, which let me know that I'd be spending my next overnight in this hotel. Forget the ship! Other than the cool bed, the room was quite normal; well, almost. The drawers were embedded in the walls. Buttons on the sides of the drawers opened them. A lamp on an ordinary-looking table with a single drawer was giving off its light by fluorescence. A clear crystal was suspended above a small fluorescent pool, emanating a beam of energy into it. The energy excited the fluorescent liquid, which then illuminated the room. It was quite ingenious. Other than this, the remaining furniture was quite normal. In the bathroom, I saw the typical items one would expect to find. The only thing that stood out was a very ornate bowl that didn't seem to fit the bathroom at all.

Above the set of drawers, I saw a large picture of a spiral-shaped musical instrument. YES! Here was another reference to that cool machine. I hoped that it was in this building. Now, because I'm nosey and I like to play with things, I checked out the drawers that were in the walls. There were five drawers in total. Checking the top drawer, I found a bible left by the Gideons. Man! Those Gideons sure get around.

No! I'm only kidding with you. I couldn't resist. In reality, there wasn't anything in the drawer. Checking the second drawer, I found an article of clothing someone left behind. By the look of it, the person left it behind on purpose. Nasty! The third drawer had a key within it. Now, I was getting somewhere. The key was about 3 inches long, with a rectangular shape at the end of it. The rectangle had a hole in its middle. *Am I to expect keys in the shapes of the buildings outside?* I thought. I couldn't wait for all of this to start making sense. At least this key wasn't inside a metal box. Storing the key with the other one I found earlier, I continued with drawers 4 and 5, both of which were empty.

Finding nothing more, I exited room A. One thing I noticed was that the key from the pyramid building and the rectangular key found in the drawer were of a different make. The pyramid key was about an inch longer than the key from room A. Continuing on, to make sure I wouldn't miss anything, I did a quick once-over inside room B. I figured it would be smart to be thorough. Looking around, I saw that it was identical in every way to room A, except for the minor damages seen in both rooms. Checking all the drawers in the wall, I found nothing at all. Apparently, room B had no key to offer me unless it was hidden elsewhere. Checking the rest of the room, I found absolutely nothing. I was about to leave when I saw the little table with a drawer in it. Opening it, I beheld a little piece of paper with three circles on it. The third circle was colored in. On the other side, I saw the Dynir characters for the numbers 1-1-8. Great! Another riddle!

Fearing I may have missed something in room A's little drawer, I went back to check it out. It turns out that there was no need to return, but I had to be sure. By finding two different items in two different rooms, I felt it would be wise, although annoying, to check out each room. Leaving room A again, I went over to room C. To my surprise, it was locked up tight. I went over to room D and found it locked as well. *Were the first two rooms a fluke?* Checking room E, I found the door unlocked. *This is pretty weird,* I thought.

Entering the room, I found it identical to rooms A and B. Checking all of the drawers; I found another slip of paper in the little table. This time, it was the first circle that was colored in. On the other side of the paper were the numbers 5-2-9. After storing the paper, I left room E. There were no more unlocked doors on floor one.

Heading to the second floor of rooms, I didn't find an unlocked door until I got to room K. After entering the room, my eyes opened very wide as I stared at a small fire. The little table that holds the fluorescent lamp had fallen over due to a damaged table leg. The collapse of the table must have been very recent; a delayed reaction from damages caused in the attack. The liquid was ignited by the crystal sphere on impact with the floor and was burning the floor and the table. I had to save the table should there be a slip of paper inside of it. Saving the hotel was a good idea as well.

Rushing around frantically in search of something to put out the fire, I remembered the fancy bowl in the bathroom of room A. Trusting that all of the rooms had one or something similar, I ran to the bathroom. Yes! There it was. I picked up the bowl and placed it in the sink to fill it with water. Carrying it to the fire, I dumped the water on it. I had to repeat the process three times before I fully extinguished the flames.

Checking the table, it was plenty charred in some places, but I was able to check the drawer, which remained undamaged. Inside, was another piece of paper that I retrieved unburned. On the paper, I saw the second circle was colored in. The other side of the paper had the numbers 0-5-3. I wondered why the number was represented as 053 instead of just 53. I had a good feeling that I wouldn't be finding any more paper slips. I had three papers, each with three circles and a different circle shaded in on each one. Doing the only reasonable action, I placed the papers in order by their shaded circle. In doing so, it made the numbers display 529053118 when put all together. The number was strangely familiar to me. Finally, it dawned on me that it was a coordinate. That would make sense as to why the number 53 was written the way it was. I hoped to find out where it would take me before using it. I wouldn't want to fly into Uganthos.

Continuing on, the next unlocked door was room N on the third floor. This time, it was a key that I found, which was cubical in shape. I found the key inside drawer number two. It was getting rather hard to keep track of all the details. In room A, the key was in drawer three, and in room N, it was in drawer two. I found papers in rooms……

Wait a minute! Of course! That's why the number seemed strangely familiar. I've seen it already, displayed near the mural in the cylindrical building. It's the coordinate for the landing site on Varyn. Excellent!

Feeling hyped up, I left room N and proceeded to room P, seeing that room O was locked. Finding room P opened threw me off a bit. I thought I saw a pattern forming in the choice of rooms that were being left unlocked. Up till now, the lettered rooms matched the letters in Varyn Keb's name. I had the A and N for Varyn, and all of the letters for Keb. What was more; the K, E, and B rooms all had slips of paper in them instead of keys. It was the A and N rooms that had the keys. But now, having room P opened doesn't fit my theory.

The first odd thing about room P that caught my attention was the fact that the door was already slightly opened. With a number of reasons going through my head as to why this might be, I very cautiously entered the room. I slowly checked out each area of the room but found no signs of any other person. Perhaps the door was opened during the attack on the city; after all, there was a fair amount of debris lying about. On closer inspection, I noticed that there was structural damage done to the far left wall. In all reality, it could even have been Noks rummaging through the rooms, and he simply didn't close the door all the way as he left.

Whatever the reason, I found nothing out of the ordinary within the room. I checked all the drawers and came up empty-handed. If anything, finding nothing was out of the ordinary! Why should there, all of a sudden, be no key, paper, or anything else for that matter? That didn't make sense! Being a bit confused, I began making my way back to the door. About two-thirds of the way there, the floor crumbled beneath me. Half of my body fell through before I caught hold by my arms on a steady area of the floor. It wasn't easy to get myself in a position where I could hoist my body back up, but I managed in the end. Rising carefully to my feet, I quickly left room P. I now understood why room P didn't fit the pattern. I was never meant to enter into it. The room was not supposed to be unlocked.

Seeing that my idea was showing credibility again, I expected room R to be the next room to be unlocked, and I was right. As I began opening the door, I was startled when the door ran into something almost immediately. The door didn't even open enough for me to squeeze through it. I really do need to cut back on the bagels. This was the second time I couldn't pass through an opening.

Although I couldn't walk into the room, I was able to see inside the room a little bit. I saw that the back left-hand corner of the wall had been partially blown away, leaving a sizable hole. The damages affected the back right-hand corner of room S. It seemed that this area of the building was the point of greatest damage. I couldn't quite tell what was blocking the door, but I did notice that some of the ceiling was missing. I, therefore, assumed that the obstruction was part of the collapsed ceiling resting against the door.

Well, this was not good. I needed to get in there. Pacing a bit while thinking of how to get into the room, I caught sight of room S's door. It was closed, but not all the way, reminding me, unfortunately, of room P. I checked the door, and it was locked. Seeing it up close, I could tell that it was slightly ajar. Looking through my gear for a tool, I took out my crowbar. Using it on the door, I broke it open, allowing myself to get through. Due to the hole in the back wall, I was able to pass through into room R after all.

Once inside, I carefully made my way around a great amount of debris. I saw that a large piece of the ceiling was indeed blocking the door. I wasn't going to get passed that door any time soon. Looking through the draws, I came across the spiral-shaped key, this time in drawer number five. It resembled the keys found so far from the other rooms, measuring about three inches in length. While storing the spiral key, I stopped to ask myself why the key that represented this building was in room R and in drawer five. It seemed like a good question, but one that I unfortunately had no answer to.

Not long after, I remembered the stone tablet in the lobby. It had the letter R on it, and the spiral key was located in room R. This couldn't be a coincidence, but as for why the fifth drawer, I was completely stumped. Looking through my notes, I found the stone tablet with the letter Y within the pyramid building. If my theory is correct, I should find the pyramid key in either drawer one or four of room Y. Room A had the rectangular key with a hole in the middle. Perhaps that key lies within what is left of that building.

Noks! He was searching through the octagonal building's remains last night. I wonder if he found anything?

Exiting room S, I made my way over to room V on the fourth floor. To cover my bases, I made sure to check all the other rooms on my way over to it. As I expected, all of the other rooms were locked. Happily, I found room V unlocked. I was now certain that the unlocked rooms spelled out Varyn Keb's name. Expecting to find another key, I looked through the drawers. This time, the key was in drawer number one, and it had the shape of a sphere. Feeling pretty charged, I left room V and headed for room Y. As I approached, I took a look over the edge of the spiral staircase. The view was rather striking, but it wasn't anything I would recommend to a person who is afraid of heights.

When I reached room Y, the door didn't open right away. For a moment, my eyes opened very wide in astonishment. Before I could whine and complain too much, the door opened up. It must have been momentarily stuck, being another side effect of the attack. Once inside the room, I went straight for the drawers in the wall and opened drawer four. There it was, the pyramid key, just as I expected. Mission accomplished! I now had all five keys and the coordinates for Varyn.

When I exited room Y, I noticed a small alcove after room Z. It connected the fourth floor to the large pillar that the stairs went around. Looking closely at the area, I saw that it was cylindrical in shape, and that there was another apparent door. Only this time, there was a keypad to the right of the supposed door with the numbers one through nine. Under the keypad were five geometric shapes, the same as the keys I just finished finding. The picture displayed a sphere, then a rectangle with a hole in its middle, a spiral, then a pyramid, and finally, a cube. Looking at the order in which the shapes were displayed, I did a little backtracking in my notes. *In which room did I find the spherical key,* I thought. I found it in room V. My notes also showed that I found the rectangular key with the hole in its middle in room A. Of course! The keys are in the order needed to spell Varyn.

Okay, that part is solved. Now for the number to input into the keypad!

Slapping myself in the head, I looked in my notes for the drawer number I found each key in. Assuming I made no mistake, the spherical key was found in drawer number one. Looking up the remaining drawer numbers for the keys in the order in which they were displayed, I ended up with the number 13542. Inputting the number into the keypad lowered a circular part of the alcove's ceiling, which I thought was just decoration. Getting out of the way of the descending ceiling, I waited for it to come to rest. When it did, a door opened, beckoning me to get inside. It was an elevator that came from an unknown room above me. There was no apparent way to know that any other floor existed. There were no stairs or any other means to get to the mysterious room. Apparently, the only way to get there was by the elevator in front of me.

Wait a minute! There was a way to know that the hidden floor existed. Back in the Rhombus building's 12 o'clock room, which represented the spiral building, there was a diagram that showed how to find the instrument. It showed its location to be the top floor of the spiral building. The lines representing the room were dotted instead of solid because of its hidden location. I didn't understand that at first, but now I do. It makes perfect sense now.

Feeling pretty good, I entered the elevator. The door closed behind me and automatically took me upward. The trip only took about ten seconds to reach the hidden room. When I exited the elevator, I stood looking at a very spacious room, about three stories in height. My attention was only stolen away by the sound of the elevator door closing and the lift sinking back into the floor. I was now apparently stranded on this mysterious floor. I did not wish to concern myself right away with my apparent incarceration. I had something else grabbing my full attention.

There it was! That incredibly cool instrument I had hoped was real. It was real alright; all three stories of it. It appeared as a very long tube that spiraled around the axial pillar. Also encircling the pillar was a walkway that would enable me access to the tube. Drilled into the tube were twenty-six holes that were equidistant from each other. From every drilled hole, there was another tube that shot horizontally into the axial pillar. I stood there wondering why the instrument's tube connected with the central pillar.

At the base of the pillar were two buttons, one lit and the other dark. From the time I stepped out of the elevator, I heard a triad being played at a pleasant volume. Examining the instrument in greater detail, I found the source of the triad. Filling three of the drilled holes were three other keys, just like those I recently found in the rooms. These keys, when inserted into the tube, created the sound that was being played. The mystery of why the tube was connected to the pillar was solved. The pillar was apparently a huge air pipe, providing the instrument with the air it needed to produce its sound. The system was backward though. The instrument required the keys to be inserted in order to produce its music. Very interesting!

Removing the keys stopped the triad, leaving the room in silence. When I pulled the keys from the instrument, I saw that they had the shapes of a cylinder, an octagon, and a rhombus. This made a total of eight keys. I was quite uncertain what needed to be done. I had eight geometrically shaped keys capable of producing musical chords, but I had no idea what I was supposed to create.

Going over my experiences again, I tried to think of any clue I may have received that was slipping my mind. Most of my clues were given in the rhombus building. It was almost like the city's library. There were a number of rooms within the rhombus structure, most of which I had no desire to revisit. If there is a clue that has slipped my mind, it most likely came from the spiral building room. On its door, it had a drawing of the instrument which stands before me. I had a frustrating time navigating a horrible maze within that room, but in the end, I found a message in a metal box that read…; Hold on, I have to look through my notes again. Ah, yes! It said to play the founder's song.

What is the founder's song?

Just a minute! Where were the keys positioned in the instrument?

I took the keys out of the instrument without really paying attention. How careless can I get?

Let's see! Of course! There are twenty-six holes in the machine. Each one must represent a letter of the alphabet. The founder has to be none other than Varyn. The five keys from the rooms were found in the rooms lettered V, A, R, Y, and N. That must mean that the keys found in the machine represented his last name, Keb. That being the case, the keys must have been in holes 11, 5, and 2, which are the positions of the letters K, E, and B.

I needed to put the keys back, not just the three initially found in the machine, but all eight of them. This is how I am to play the founder's song.

Beginning at the first hole representing the letter A, I placed the A key in the machine. It took a moment, but I realized that the instrument wasn't giving off any sound. It was playing with the three other keys within it. Why was it not working now?

I then got the idea that perhaps the keys had to go into the machine in the correct order. It seemed a bit odd, but it made sense when you think about it. After all, Varyn doesn't start with an A. Testing my hypothesis, I walked all the way up to position number twenty-two, representing the letter V, and inserted the key. A tone emanated from the instrument. Yes! That was the answer! Although a bit annoying, I then returned to the first hole for the letter A and inserted the key. Tone after tone was heard as I went about inserting the keys. Listening to the sound grow with every key was an enriching experience. The final masterpiece that emanated from the instrument when all the keys were inserted was not easy to describe. I was certain the final result would be dissonant, but the complexity of what I heard went far beyond what I was used to hearing. Although complex, its harmony was unmatched. It was a beautiful sound that felt like it expanded my mind onto a higher plane of thought. I don't know how else to describe it. Even still, my words do not do it any justice.

I listened to the majestic sound for a bit longer, wondering what needed to be done now. As I looked around the room to see if anything was different, my eyes fell upon the buttons at the pillar's base. The button that was dark had become illuminated. Curious to know what the new button would do, I pressed it. Slowly and sadly, the incredible sound faded away. I apparently shut down the instrument's air flow. Moments after the room became silent, I heard a sound like a door sliding open. Looking around, I couldn't see any evidence to support what I heard. I knew that the sound came from above me, so I went up the walkway as far as it would take me. From my new position, I was a little bit higher than the pillar and was able to see that the top of the pillar had opened. Walking over to the opening, I saw that it was another elevator. The pillar had a duel purpose. I wondered how the pillar could provide sufficient air for the instrument with an elevator jamming the pipe. While entertaining my thoughts, I saw that the keys were still inserted in the instrument. Not wanting to leave the keys behind, I retrieved them again and placed them in my backpack. Once stored, I returned to the elevator within the pillar and got into it.

Once inside, the answer to my earlier question was given as the elevator descended. Looking down at my feet, I saw that the floor of the elevator was capable of opening up. No doubt, when not being used as an elevator, the floor remains open to allow unrestricted air flow. I hoped very much that the elevator floor would not malfunction. Desiring to focus my attention away from the floor, I noticed a beautifully decorated box on a small shelf that had managed to escape my view. Inside was the spiral-shaped key. Like the pyramid key found earlier, it had the same beauty, length, and make.

I now had two keys in my possession. Things were coming together slowly but surely. When the elevator door opened, I saw that I was staring into the lobby. It was not the ceiling door that opened but a door that was in front of me. When I exited the elevator, which I did very quickly, I saw that it was the door I initially found on the pillar that seemed to have no way to open.

Looking things over to make sure I didn't miss anything, I came up with nothing new and decided to leave the spiral building. Except for almost falling through the floor in room P, I enjoyed the experience the spiral building had to offer. It was a lot nicer than any others I had faced. Breaking free of my thoughts, it was time to move on, but I told myself that I'd be back when it was time to hit the hay. I hadn't forgotten that floating bed.

CHAPTER 27
REMNANTS

While leaving Varyn Estates, A.K.A, the spiral building, I saw that I was about to enter the area of the city that was hardest hit during the attack. Three of the next four buildings were nothing but rubble strewn about. From the information I gathered earlier, I was heading for what used to be the octagonal building. When I first beheld the city, this was the only structure I was totally clueless about when it came to its shape. If not for the information I found in the pyramid building, I still wouldn't have a clue; it was that badly damaged. The most intact area of the building was a piece of its outer wall that still stood about nine feet in height. I suppose if I had seen the remains of the structure from this vantage point, I might have had a better guess that it was once octagonal in shape. I say this because the outer wall stood atop a piece of wall that protruded from the structure's base at a forty-five degree angle relative to the ground. This made all the inner angles measure one hundred thirty-five degrees. Continuing this pattern would eventually make an octagon.

The rubble I saw before me ranged in size. The smallest pieces could be moved by hand, whereas the larger pieces could only be moved by a machine. It was a depressing sight. The vast majority of the building lay in ruins. There was furniture, artwork, and other items scattered all about. Richard did not say where he hid the keys, so, unfortunately, I needed to sift through this mess. Rummaging through what I could with my hands, I found wonderful things that were either damaged or destroyed, but nothing like a key. With a high probability that a key or some other needed item could still be buried, I went back to the Lotania and retrieved the Land Rover. Hopefully, its robotic arm would be able to move the larger debris.

The task of searching through the octagonal building's remains took about two hours to complete. When I got done moving what I could, all I came up with was the octagonal building's stone tablet. The Dherney character on the tablet was the letter E. You know,....I think I got close to pronouncing the name that time. I'll get there.

Leaving the sad remains, I drove the rover to the next heap on the ground. In some places, it was hard to tell where the octagonal building's pieces ended, and the next building's pieces began. The next structure I'm referring to was the rectangular building with a hole at its center. Although destroyed, it had a bit more of its structure still intact. It was easy to see the inner and outer walls, which gave the clue that a courtyard most likely existed here. Wasting no time, I began moving large chunks of debris with the rover. I unearthed many items that I will not list.

Having a bit less debris, I was able to search the area in about one hour and fifteen minutes. When I completed my task, I had found two items of interest. One was the building's stone tablet, which had the letter A on it. This made sense, seeing that the short rectangular key was found in the drawers of room A in the spiral building. The other item of interest was a metal box that contained, can you guess? If you said a key, you are absolutely WRONG AGAIN. No! Inside this box was another message, written, surprisingly enough, in English. The message read:

"Overcome them with darkness."

I had three questions: who was being referred to as them, why was the message written in English, and what did it all mean? I couldn't help but be interested in these questions. Unfortunately, as usual, I had no answer to give myself, yet.

Ready to move on, I drove the rover to the last destroyed structure. This meant having to temporarily skip the spherical building that followed the rectangular structure. Somehow, the spherical building managed to escape destruction, similar to how a house stays standing even though its neighbors get destroyed in a tornado. I would return to learn the sphere's secrets shortly. For the moment, the remains of the cubical building needed to be dealt with.

Just like the last two heaps of rubble, I needed the rover to find whatever items could exist, if anything at all. Searching through the debris proved to be a bit more challenging than I expected. Many of the pieces were too large for even the rover to move. Others got pushed into the dirt under the force of stones that fell on top of them. It took me a good three hours and twenty minutes to search the area. Even still, I'm sure that I didn't clear everything away due to the conditions I mentioned.

I was able to find two items of interest that satisfied me greatly. One item was the expected stone tablet. I almost overlooked it because it was lying face down in the soil. Flipping it over, I found the letter N carved into it. The second item was the prize winner. Inside a metal box which was under deep rubble, I found the cubicle key. It was ornate and four inches long; it was truly a beautiful artifact. Having three of the five keys, I felt things were going rather well. Checking the time, I found that it was 46:89:72 rotations. I still had time, but not to rest. Off to the spherical building!

CHAPTER 28
ALIEN PATTERNS OF LIGHT

Before entering the spherical building, I returned the rover to the Lotania. A couple of minutes later, I was looking at a remarkable work of architecture. The inside of the building was completely spherical, except for the floor, of course. Even still, the floor was round in shape. As for the décor, there was nothing in the room except for a computer that was built into a table. The computer was equipped with a monitor, controls, buttons, and lights. The colors of the lights were red, yellow, green, and white. I noticed that the red light was lit. The monitor itself was inclined at an angle. Attached to the computer was a device that I wasn't completely sure about. I had my ideas of what it looked like, but at this time, I'll keep quiet. The area in front of the monitor was flat, with two items lying on it. One of the items was a star map with nine constellations drawn on it. I recognized the constellations as being the same ones displayed within the rhombus building inside that horrible rotating room. I get the heebee jeebees just thinking about that place.

The other item was a set of nine micro discs lined up neatly on the table, with the fifth disc missing. The discs were very interesting. They weren't flat like the old CDs of decades ago. Instead, they looked almost like flying saucers. I assumed that they were storing information that I would soon learn about.

Below the green and yellow lights was a small featureless compartment that I could not open. Located under the lower left corner of the computer screen, I saw a long, skinny rod sticking out of the computer. At the end of the rod was a small indentation capable of holding a micro disc. I wasn't quite certain what needed to be done at the moment, but I was certain it had to do with the computer and micro discs.

Looking around the room again, I saw no other items. I did happen to notice that the walls had a lot of tiny holes within them. At first, I didn't think much of it. I figured it was just the texture of the wall. But looking at it much closer, I saw that the holes didn't cover the dome in all places. There were holes in a particular area and then a smooth area, followed by more holes. It seemed that the holes were randomly placed around the room.

My idea about the device attached to the computer was making more sense now. I was thinking that it looked like a laser, and I just might be right. Still uncertain of what to do, I head back to the computer. The only sign of life on the computer was the red light. Even the monitor was blank. Out of curiosity, I touched the red light.

Uh oh! What did I do? The main entrance door began to slowly close, and the room lights slowly faded off. As the room lights faded off, a red light across the top of the computer table slowly turned on. Over my head, the ceiling and walls came to life as the night sky revealed itself. As my eyes adjusted to the darkness, I could see the stars much more clearly. The sight was overwhelmingly breathtaking. I understood now that the holes made constellations of various items. Other areas on the dome were just black patches. I tried to recognize some of the star patterns on the map, but I didn't do too well.

Looking down to rest my neck a bit, I noticed that the monitor woke up. It had a message on it written in the Dhorniy language. I was getting better at reading the language but still accepted the aid of the decoder here and there. The message read:

"Insert a disc."

As I reached for one of the discs, I saw that the yellow light was now illuminated and that the red light had turned off. Touching the yellow light caused the micro disc arm on the computer to retract. A few moments later, another message appeared on the monitor that read:

"No disc."

Touching the yellow light again, the drawer came out, and I placed a disc on it. I randomly picked disc number seven. I had no particular reason. Touching the yellow light again, I retracted the arm and disc into the computer. Moments later, an image of a spiral pattern came up on the monitor. Underneath the image was the Dhirnie word, which read:

"Completed."

That must be it! The disc displays a constellation that gets drilled into the dome by the laser. This particular disc for the spiral pattern was already drilled. Trying the other discs, I ended up with the same results until I put disc number two into the computer. On the monitor, an image of an octagon was displayed. The laser drill aimed itself at an area of the sky that was dark. The yellow light shut off, being replaced by a green light.

I was sure that by pressing the green light, the laser would create the pattern of the octagon on the ceiling. The night sky was being created on the dome above me. This had to be the oddest planetarium I'd ever visited. Activating the laser by touching the green light, the laser did its job with high precision. It was a hypnotic display of blue beams that shot up to the ceiling. They were almost too bright to look at. I watched as star after star of the octagon pattern was drilled into the dome.

The process took about six minutes or so. When the laser stopped, I saw the word completed displayed on the monitor. The yellow light turned on again, replacing the green one. Of all the discs on the table, the only one left that needed to be drilled was disc number nine. When loaded into the computer, an image of a cube came up. Again, the laser aimed itself at an unfinished area of the ceiling. The green light returned, waiting for the command to commence its drilling process again. Allowing it to do so, the laser created the cubicle pattern over my head. The process only took about four minutes this time because the pattern was simpler than the octagon.

When the pattern was completed, the red and green lights were illuminated instead of the yellow light. Also, instead of the word completed being seen, a message which read:

"Do you wish to create the Varyn pattern?" was displayed.

Having no traditional yes or no buttons to respond with, I had to assume the red and green lights to be the equivalent. I had to see this project through, so I touched the green light to indicate yes. Apparently, I touched the correct light to proceed, but instead of the laser moving into a new position as I expected it to, the small compartment under the green and yellow light opened up. A moment of excitement went through me as I looked upon the disc that was missing. I finally recognized that the yellow light had illuminated again, turning off the red and green. Touching the light, the retractable disc tray came out of the computer. I removed the disc still on the tray that held the cubicle pattern and replaced it with the missing number five disc.

Touching the light again, the computer loaded the disc and displayed the image of a person, presumably Varyn. With the image on the monitor, the laser positioned itself for the creation of the new star pattern. When the laser came to rest, it began firing its beam at the ceiling without the green light needing to be touched. The pattern was one of the most detailed of the nine. It took close to ten minutes before the laser stopped. On the monitor, a new message read:

"Do you wish the final star completed?"

Within the image on the monitor, the star that sat in the center of the image's chest was blinking. I looked at the star map on the table, and there it was: the constellation's brightest star. Again, the red and green lights were back. I touched the go-ahead light to complete the Varyn figure. The laser fired a prolonged beam at the chest area of the Varyn pattern. When completed, I could see that it wasn't only the brightest star of the Varyn constellation, but of all the constellations that had been created.

I then saw a beam of light coming through the newly created hole representing Varyn's chest star. The beam came down at an angle like the one from the rhombus building. It reached the floor a little way beyond the computer. In fact, it touched the floor at the very center of the room. Not long after the beam was created, I found myself panicking as I floated slowly upward toward the beam. Somehow, the force of gravity became null and void. I had no idea what to expect upon making contact with the beam. I couldn't help feeling that I might be vaporized shortly.

Soon, I found myself entering the beam of light. Thankfully, instead of vaporizing me, it was like being caught in a tractor beam that pulled me. Only in this case, I was pushed. Gravity was slowly being restored as I began to descend to the floor at the angle the beam was following. I was heading for the center of the floor. As I continued to descend, a square made up of nine smaller squares illuminated the floor. What did I have to do now, play Tic Tac Toe?

Within each square was a constellation from the star map. When I touched down, it was in the center square. Immediately, the surrounding squares started shuffling their star patterns. The center square where I stood did not change. The squares shuffled for about five seconds before stopping. When I stepped to a different square in my attempt to exit the main square, I noticed that the pattern I was stepping on moved to the new square I had stepped into. All of the other patterns shuffled again. What was going on?

I was obviously dealing with another puzzle. I could only guess that the constellations had a certain order they needed to be in. Other than the star map, I had no clue of any order. I, therefore, retrieved the map from my backpack and saw that the patterns were arranged in nine blocks, like the square pattern on the floor. Having the answer to the puzzle, I was ready to solve it. From what I had experienced already, stepping on a square kept it from shuffling, locking it in place until it was stepped on again. I could therefore move the patterns from one square to another by stepping where I wanted them to go.

Looking at the star map, I began putting the star patterns in their correct order. When I succeeded in positioning all the star patterns correctly, the eight outer squares within the main square started blinking. The center square became exceedingly bright and, a few seconds later, rose up out of the floor, revealing a compartment in which a metal box resided. It is needless to tell you how much I was smiling at the time. As the square rose out of the floor, the main lights slowly came on, and the door to the spherical building opened. Inside the box was the spherical key I was hoping for. Just like the other keys, it was artistically beautiful. I carefully stored the key in my pack and made my way outside.

Chapter 29
The Domed Central Complex

Exposing myself to the light of a giant orange star was a bit challenging at first due to having spent so much time in the dark. In all reality, Uganthos' light was incredibly intense in comparison to the Earth's Sun. Even so, being an orange star, it made my surroundings seem a bit darker than what I was used to. It didn't take me long for my eyes to adjust to the daylight.

OK! There was one building left to examine and one key left to find. Hoping to finish my search for the keys, I made my way over to the domed central complex. Unlike the other buildings in the city, I found the main entrance to this building locked up tight. Having the technology it had within its walls, I could understand why. Looking for a way in, I noticed to the left of the door that there was a small square indentation on the wall. If I had approached this building before investigating the others, I probably would not have known what the indentation was. But having visited the other structures first, and collecting all the artifacts I have, I was able to identify the indentation to be a keyhole. It also means that I would not have been able to enter this structure due to the lack of a key.

The key that came closest to matching the square indentation in the wall was the cubicle key. After placing the key into the niche, the door to the central complex slid open. Taking a few steps into the building, I stopped and returned to the main entrance to retrieve my key. I wasn't about to lose track of them after what I went through to get them. Once inside, I entered into a small foyer with art decorating its walls. There was also a small square table with a surface that was illuminated from within. Beyond the table was another door.

Looking closely at the table, I found another message which read:

"Place your keys and identify your rank."

Below the message was also written:

"One key, two keys, three keys, four keys, and five keys."

It seemed that I had a multiple-choice question on my hands. The answer was obvious that I had to choose the four keys option. After placing my keys on the table, I touched the four key choice. A moment later, the lit tabletop began to blink. About three seconds after that, the table went dark, and an alarm began to sound. In a state of panic, I ran for the main entrance. Before reaching it, I stopped again and went back for the keys. *I'll get the hang of this key thing eventually.*

Running out of the building, I found a place to hide myself. Breathing heavily, my thoughts became more stable after a few moments. *What did I do wrong, and WHY did I run?* The city was deserted. There wasn't anyone to apprehend me. I felt rather silly after thinking about it. Of course, I could have been gassed, or lasered, or something. I couldn't afford to take any chances.

Not long after hiding myself, the alarm stopped blaring. It either shut itself off somehow, or someone else was still around to shut it down. *Could it have been Noks?* I thought. If so, he wasn't coming out of the building. I stayed put for a bit longer to make sure the way was safe. When convinced enough, I slowly approached the building again. All seemed as though nothing had ever happened. I, therefore, acted as though nothing had happened. I placed the cubicle key in its indentation, and the door slid open again. *So far, so good!* Staring at that dreadful little table, I noticed that its top was illuminated once again. Approaching the table slowly, I saw the messages displayed just as they had been before.

What did I do wrong last time? I thought. Thinking things over for a little while, I then asked myself aloud:

"Who is the owner of these keys?"

I never thought of them before as personal property, simply as random objects. The last person to have had the keys was Richard. He said he had them with him at the time of the attack and that he hid them away for safety. The keys must, therefore, be his. *But why would he own keys here*? I thought. *This is very odd! If the keys are his, what would he need them for, and what is his rank?* Thinking for a few seconds, I found myself nearly slapping my head again. This was a DUH moment. He had all five keys in his possession when he hid them. His rank must, therefore, be five.

Getting myself ready to grab and run again if necessary, I held my breath and placed the keys on the table. This time, I chose the five-key answer. The table started to blink again, and every muscle in my body tensed. After three seconds, the blinking stopped. To my great relief, no alarms went off. Instead, the computer displayed another message. The translation read:

"Axnard Jirroch, Nirton Zesaq, and Caynos Quosh."

Having no idea what these words were, I had to assume they were names. Perhaps they are the three possible individuals that could be the owner of the keys.

Wait a second! I had heard one of these names before. Rich told me during one of his holo-speeches about an Axnard Jirroch. He said he was the president of their people.

Well, it was time to take a guess as to which name was correct. Like before, I got myself ready to grab and run should it become necessary. To make a long story shorter, I did need to grab and run once. In the end, though, I found out that the name associated with the keys was Nirton Zesaq. I could only surmise that the three named individuals were people of authority, having the highest rank available. It didn't tell me how Richard ended up with the keys. I guess it was something I'd need to ask him when I see him again.

When the correct name was chosen, the door at the end of the foyer opened. Picking up the keys from the table, I cautiously entered into the next room. It was an incredibly spacious room with a height of about forty feet. The room was lightly decorated with statues, pictures, artifacts, and murals. The area was adorned with a number of pillars that went all the way to the ceiling. It was quite impressive, to say the least. Unfortunately, parts of the ceiling caved in, leaving a good deal of debris lying about. The floor had a number of colored tiles that made it up. It wasn't easy to see what the colors were all about due to the debris.

On the opposite side of the room from where I entered, I saw very large metallic doors, which resembled those of an elevator. The doors were most likely the entrance to the domed area of the structure. They looked as though it would take a nuclear blast to blow them open. I assumed the time-altering device lay just beyond those doors. In front of the doors, I saw a podium of sorts that stood about four feet in height. On the top of the podium, I noticed five indentations. One of them was for a spiral key; the indentation looked like the letter C. One was for a pyramid key, the point needing to go in first. Another was for the spherical key, which had a bowl-shaped indentation. The cubicle key with a square indentation was to the right of the spherical indentation. And finally, the rectangular key with the hole at its center had a rectangular indentation with a raised middle, much like that of a mesa. *It will take all five keys to open this door, which means it won't be opening for me any time soon.*

Continuing to look around, I noticed a catwalk high above the floor. Under the catwalk was another podium, this time with four indentations. It consisted of the same indentations as seen on the podium near the massive doors, minus the rectangular key. The catwalk itself was a straight path about twenty feet long and three feet wide. The path started at one of the pillars and moved away from it until it reached a wall with something on it. Moving about the room, I saw that the path continued beyond the wall, leading to a platform containing a table with a metal box on top of it. The wall extended about three to five feet beyond the catwalk on both sides of it. I could see no visible way to get around it. Going over the wall was also out of the question. The wall extended all the way to the ceiling. I had to get up there to figure out what was on the wall and how to get beyond it to reach the metal box.

While walking around the room to check out the catwalk, I noticed that the pillar where the catwalk began had rungs embedded within it. Although sturdy, the pillar was showing signs of having received damages due to the attack. A couple of rungs were damaged. One had come loose, dangling from the pillar, and another had fallen out completely. Thankfully, the two damaged rungs were not near each other. This stroke of good luck allowed me to eventually climb the pillar. I say eventually because as I climbed the pillar, I very luckily caught sight of a shimmer or distortion as I looked up at the pillar. It was not easy to see, but as I stared at a solid object up toward the ceiling, a distortion, much like looking through heat waves, became visible. Fearing to climb any higher, I returned to the floor to test my theory. I took out my grappling hook gun and aimed it at the catwalk. This was the way I originally figured I would be getting to the catwalk, but when I saw the rungs, I just figured I'd climb. I could use the exercise. Remember my bagel problem?

Firing the gun had interesting results. As the grappling hook went up, it encountered whatever the nearly unseen barrier was. The hook violently bounced off the barrier as if fired from the gun a second time. The force nearly wrenched the gun right out of my hands. It seemed like an electrical discharge when the hook met the barrier. If it had been me that made contact with the force field, I'm sure I'd be quite dead right now. It would have blown me clear off the rungs to the floor below me. That was too close a call and something I chose to forget.

The answer of how to proceed was quite obvious, and I don't know why I didn't take care of it right away. The podium underneath the catwalk wanted four indentations filled. I had the keys in my possession. I don't know why I didn't just fill them in. Perhaps I thought I was missing one of them. *I don't know! I must be getting very tired again, causing me not to think clearly.*

Installing the four necessary keys didn't appear to do anything, but when I shot the grappling hook at the catwalk again, this time, it latched onto it. The podium was the controller for the force field. Had I installed the keys in the first place, I would have avoided possibly getting a permanent haircut.

With the hook latched to the catwalk, I retracted the cable and took a free ride upward. When I was within arm's reach of the catwalk, I grabbed hold of it and hoisted myself up the rest of the way. Once on the catwalk, I carefully made my way over to the wall. As I figured from the ground, I could see no way around or over the wall.

What I did see was an area of geometric indentations within the wall, making a rather symmetric and complex pattern. Below the geometrical phantasm, I found thirty-three pieces of different colored tiles, matching the smaller indented shapes within the main pattern. The pieces were held inside a half bowl-shaped holder, which was attached to the wall. It was obvious what needed to be done. The real question was how to do it. Many of the tiles were the same shape but with different colors. I couldn't just randomly place tiles in the pattern. I needed to figure out where I may have seen the answer to this puzzle. Then I remembered that messed up experience in the pool. The parchment inside the tube that got damaged had colorful geometric shapes on it. That paper had my answer to this blasted puzzle, and now it's lost for good. Fantastic!

I was not happy with the idea of having to guess my way through the intricate problem. Quite frustrated, I turned around and looked across the room from my impressive height. It was actually very stunning, momentarily taking my mind off the problem at hand. I'm surprised I took as long as I did to look at my surroundings. I will say that this was no place for someone with a fear of heights. Thankfully, I did not have such a fear. Still, looking down was a bit unnerving, and I didn't plan on repeating my action a second time.

Just then, I found myself taking back what I just said about looking down. From the floor, I could see that it was comprised of many colors, although I was unable to make anything out of it. From up here, though, it was a different matter. There it was! The answer to the puzzle lay displayed beneath me across the floor.

Woohoo! This was more like it!

Some areas were covered by the debris of the fallen ceiling, but most of it was visible. When needed, I'd go down and move the rubble out of the way. Without wasting any time, I began putting together the puzzle on the wall, using the floor as my guide. It took a little while due to the complexity of the pattern, not to mention having to go back and forth between the wall and the view of the floor. What really took the most time was having to descend to the floor and clear away the debris. But eventually, I found all the correct pieces and completed the puzzle.

When done, it activated the centerpiece of the puzzle, which was white, causing it to illuminate. Touching the lit piece caused the entire wall to rise up high enough into the ceiling for me to slip underneath it. Having an unobstructed path now, I made my way over to the table with the metal box. Inside, I found a note that had the following words written on it:

"I have the key (RG)."

This message was written in English, just like the last one I found. I won't say that I wasn't greatly disappointed. I was hoping to find the last key and restore the city. Instead, I needed to find Richard to get the key. Don't get me wrong; I need to find Rich anyway. It just wasn't what I was expecting to find, that's all. Well, I guess I get to kill two birds with one stone, as the saying goes. ....You know, that isn't a very happy saying when you stop to think about it.

What am I saying? I need to focus.

Alright, where do I find Richard? In his last message, his ship was broken into by the Allorayans, who abducted him. I later saw him and one other forced onto one of the Allorayan vessels. As long as Richard wasn't on the ship that engaged me in space and then went bye-bye, he should be on the planet Allorayus. This had now become a rescue mission. It seemed I was going to Allorayus. This was going to be fun.

A couple of minutes after leaving the domed complex, I heard a beeping sound coming from my pack. It went on for about fifteen seconds and stopped. It was the timepiece I found on Rich's ship. The time read 69:20:77 rotations when I saw it. It reminded me that I needed to be on Varyn at 81:75:00 rotations. I calculated that the remaining time was equivalent to about three and a half hours. I needed to move if I was going to rescue Richard and still make it to Varyn on time.

CHAPTER 30

NOKS REVEALED

On my way to the Lotania, guess who decided to make an appearance? For the longest time, Noks was nowhere to be found. When I was in a hurry, that's when he decided to show up. The following is the conversation Noks and I had to the best of my memory. Oh, and by the way, Noks just happened to pull a weapon on me as I greeted him. Nothing major!

"Noks, where have you… what are you doing?"

"Hand over the keys you have been collecting, Captain Travis."

"Noks, what's going on? I thought we were trying to look for the keys together in order to restore time and bring your people back into phase with this time frame."

"My people! They are not my people. I would never have anything to do with this world, except, of course, to see it neutralized. I am just sorry that our attack didn't destroy everybody here. I was surprised when I saw you, having first thought you to be one of them. Not knowing who you were or why you were here, I decided to pose as a member of the Dhirniy world. At least by so doing, I would learn your intentions. When I learned of your plan to restore this city and its people, I knew you were their ally. I kept you alive to learn as much as I could about you, and to allow you the time to find the artifacts that would be an instrument in the future demise of my people. I will not allow you, or anyone else from this world, to bring about the destruction of my people. We will use the device we have constructed to destroy this world first. Even though it was unfortunate that the Dhirniy were not destroyed, it was to my benefit that all the inhabitants of this despicable world had vanished in the attack. It allowed me the freedom to discover the technological secrets the Dhirniy would eventually use to destroy my world."

"Noks, you don't seriously believe anything that you're saying. This race of people has no desire to destroy your world or any other. From what I have learned of them, they are a peaceful people, and I can assure you that I am no ally to them. Until this day, I had never heard of a Dhirniy. My people haven't even reached light travel until now."

"You no doubt achieved light travel on your world with the help of the Dhirniy people. They are trying to unite their race with yours and probably many others to see to the annihilation of my world."

"No, that's not it at all. Why are you seeing things so negatively? An ambitious colleague of mine on my world worked on building an engine capable of traveling near light speed. He eventually achieved it after many years of effort. In my colleague's travels, while testing the new engines, he stumbled along this race of people. My colleague shared his success story with me and invited me to visit. Out of curiosity, I accepted his invitation. I expected to meet my friend and the leader of the Dhirniy people when I arrived. Instead, I walked into a deserted and desecrated city. The only word I received from my friend talks of his eventual capture by your race. These people and I mean you and your people, no harm."

"How do you know so much about the key artifacts and their functions if you have received no word concerning them?"

"My colleague gave me what information he knew. Other than that, I have been collecting information while investigating the city. I reasoned it out the best I could."

"I take it, therefore, that you are fluent in the language of this people."

"I have been given the basic alphabet of these people to use in deciphering their language, but I am nowhere near fluent in understanding their language."

"How did you get the alphabet if everybody is missing?... ENOUGH OF THIS! You will need to come up with a better story if you wish me to believe you. How many of the key artifacts did you find?"

"Noks! Let's be reasonable."

"HOW MANY? I will not ask again."

"I have four of the five keys."

"Where is the fifth key?"

"I haven't found it. I would have figured that you had found something by now."

"I never tried looking for anything. I was just keeping my eye on you from a distance. Four keys are as good as no keys. Nevertheless, give them to me. I can't afford for you to find the fifth key and then, by chance, end up restoring this place. This city would be crawling with those wretched people in no time. Hand them over... NOW!"

Noks stood there angrier than a wet hornet with his gun aimed at me. He held out his other hand, waiting impatiently to receive my pack. I only had a few moments to do something before getting shot. I'm surprised he didn't just shoot me and take the bag. His hesitation gave me the time I needed to act. I took the bag off my shoulder as if to hand it over to him. Instead, I whirled it violently at his head, making contact with it.

Noks staggered greatly in a state of disorientation, dropping his gun in the process. I approached Noks and punched him dead in the face as hard as I could. Noks went down hard and fast. While unconscious, I tied his hands and feet using rope from my pack. I picked up his gun and made my way to the Lotania with Noks over my shoulder. Before reaching the craft, I changed my mind and headed over to Rich's ship. From my investigation of his ship, I remember it being a bit more advanced with a few extra gadgets. Once aboard his craft, I heard a familiar voice say:

"Welcome aboard, Captain Travis."

CHAPTER 31
A TRIP TOWARD HOSTILITY

"Is this your partner, Captain?" asked SAROS. "Why is he tied up and unconscious?"

"Partner? He's no partner, SAROS. What makes you ask that?"

"He entered the ship during your absence saying that he had a special piece of equipment that needed to be installed. I asked what its purpose was, but he said that I should ask you. He told me not to investigate it in any way until I spoke with you. I found his behavior rather odd. What is its purpose, Captain?"

"What piece of equipment are you referring to, SAROS? Where is it?"

"Your friend placed it under the command console."

Looking over at the command station, I could see a small device attached to the underside. Hurrying over to the object, I saw that it was a bomb.

"SAROS, this is a bomb. I need you to scan the device to let me know how to deactivate it safely."

"Right away, Captain." SAROS then continued, "Tell me, Captain, why would this individual set an explosive device on board this vessel? Does he not understand that it would destroy him too?"

"I'm sure it wasn't his plan to be onboard the craft at the time of its destruction. This individual is a spy from the planet Allorayus. He goes by the name Noks. He was involved in the attack on the city. He and his people were hoping to annihilate the inhabitants of the city, but instead, they damaged a time-altering device within the central complex. The damages caused the people of the city to go out of phase with normal time."

"How's the scan coming along?"

"Scan is complete, Captain! I can detect no trigger devices that will set it off if removed. It is undoubtedly set to explode shortly after engaging the ship. Apparently, he did not expect his plan to be foiled."

"Well, it's obvious that he doesn't know you well enough, SAROS."

"Nor you, Captain." Replied, SAROS.

Looking closely at how the device was attached, I could see nothing special. I grabbed hold of it firmly and pulled it off the underside of the console. With the bomb now in my hand, I noticed that a countdown had started. SAROS was right about it not going off right away, but apparently didn't consider starting a countdown.

The numbers were in the Dhirniy language, apparently used by the Allorayans as well. Without caring to translate, I ran out of the ship to get rid of the blasted thing. I ran in the opposite direction from the ship and city, out to a distance of about fifty yards. I threw the device as far as I could. Without hesitating, I ran back to the ship. About ten yards or so away from the craft, the bomb exploded, sending me to the ground. Keeping my head protected, I waited for the blast to subside. When all seemed settled, I got to my feet and re-entered the ship.

"That was too close, SAROS. Did the blast cause any harm to the ship?"

"No damage, Captain. All systems are operational."

"Excellent! Excellent!"

I caught myself staring at the control station and said, "By the way, SAROS, I've been meaning to ask you what this large red control is. The Lotania doesn't have it. And why are Rich's controls written in Dhirniy and not in English?"

"The red control activates and deactivates a cloaking device, Captain. Moving the lever forward cloaks the craft while pulling it back again uncloaks the craft."

"Wow!" I exclaimed. "A cloaking device! That is so cool. I wonder why the Lotania didn't get one."

SAROS went silent on me as if thinking of something to say. I then continued, "Now, how about my second question, SAROS."

SAROS continued to remain silent for a few more seconds. I continued again, "SAROS, is there something wrong?"

"No, Captain. There is nothing wrong."

"Then can you please tell me why Rich's controls are written in Dhirniy?"

SAROS hesitated again as if to exhale and then said, "It is because Mr. Gaber is Dhirniy, Captain. It is his native language. This is his home planet. He asked me to keep his identity concealed unless absolutely necessary. I'm sorry, Captain. I can assure you that no deception was purposely intended. Richard's true name is Nirton Zesaq. He is the President's second in command, the vice president of this city."

It was now my turn to go silent. It all made sense. I just didn't allow myself to believe the evidence before me because I worked with Rich for the past several years.

"Are you alright, Captain?" Asked SAROS.

"Yes, I'm fine. It's just a lot to take in, but it does make sense." I then added, "In that case, let's go rescue Nirton. He has something I need and a great deal to talk to me about."

"Yes, Captain. But if I may ask, wouldn't you like to repair the time-altering device before we leave?"

"I'd love to, SAROS, but in order to repair the device, I need all five keys. I currently have only four of the keys. The fifth key is still in Nirton's possession on the planet Allorayus. We must head there immediately. Please set a course for Allorayus."

"Right away, Captain. The course is laid in."

"Engage! Oh, and please activate the cloaking device and radar screen."

"The cloaking device and radar screen are now activated, Captain."

"Thank you, SAROS. And thank you for trusting me with the truth."

"You're welcome, Captain. I would have liked to have told you sooner."

I wasn't certain how to feel right about now. Part of me was excited to know that my friend was an alien from a different world. Part of me felt betrayed and untrusted, although it was foolish for me to feel that way. What would I have done if I were in his shoes? I can't see how I could have known he was from somewhere else. His obsession and eventual success with the engine he built could have given a hint to his identity, knowing all that I know now. But seeing that I didn't know anything, I passed it off as raw diligence and determination. That made far more sense than believing him to be an alien. He really passed himself off as a human very nicely. Makes me wonder how many other people on Earth may be holding secret identities.

Breaking free of my thoughts, I found myself looking at the radar screen and a small red x located on the planet Allorayus. I then asked,

"SAROS, I see that the planet is being guarded by Allorayan spacecraft, represented by these yellow dots, but what is this red x on the surface of the planet?"

"Ah! Yes, Captain. Nirton was working on a way for his ship to keep track of his whereabouts should he ever need to be found. He was planning on embedding a chip within himself that would send out a signal to act as a homing beacon. I didn't know he succeeded in finishing it. I, therefore, must assume that the red x is Nirton's location."

"Can you please take us safely to his location?" I asked.

"Yes, Captain. I am altering our course slightly at this time."

"Have we been detected by any of the ships in orbit?"

"No, Captain. We are totally undetectable to their sensors. We should have no problem with the Allorayan guards. The power emanations of the ship are cloaked as well. We will land on Allorayus in three minutes, eighteen seconds."

"SAROS, as soon as you can, please begin scanning the surface for any guards, traps, or any other possible problems."

"I will begin scanning immediately, Captain." "We will land in two minutes, thirty-six seconds."

Chapter 32
Rescuing Nirton

"I will be landing the craft behind a hill not far from a structure," SAROS said. "Nirton's signal is coming from this location. Posted around the building's base are six guards, each sixty degrees apart from the other."

"Thank you, SAROS. That helps me a great deal. While I'm out, keep yourself and the ship safe."

"Don't worry about me and the ship, Captain. Just keep yourself in one piece."

"You bet! I'll be back shortly with Nirton."

"I look forward to it. ....We are now on the surface, Captain."

I made sure to be well-armed before exiting the craft. When the ship's door opened, a hot breeze hit me in the face. I forgot that this planet would be uncomfortably warm. Although Uganthos was setting, the temperature felt like it was still over one hundred degrees. After exiting the craft, I took Noks' unconscious body and laid it on the ground in a shaded area. I then checked out my immediate surroundings. From where I stood, I could just see the top of the structure SAROS mentioned. Carefully making my way closer to the top of the hill, I could see that the building was cone-shaped. It was brilliantly illuminated against the darkening sky. In fact, it was very striking to look upon. I could already feel myself sweating after only a minute or two. I figured it was because of the temperature, but I wasn't going to rule out nervousness.

At the top of the hill was a huge dried-up tree, which I used to keep myself hidden from the Allorayan guards. From the tree, I could see that I was about forty yards from the structure. Off to my left was a decent-sized boulder around fifteen yards away. To my right was a partial wall belonging to an old structure that was about twenty-five yards distant. From these locations, I hoped to be able to get clear shots at the guards. My goal was to tranquilize them, not to kill. They would sleep for about six hours and wake up with a bear of a headache. After all, I had no real quarrel with the Allorayans. Well, perhaps a little one, with the pilot who tried to disperse my atoms all over space, and look how he ended up.

Before making any moves, I needed a plan on how to proceed. Once I started my assault, my cover would be blown. The tree was built nicely for shooting and for hiding. Its massive trunk went up about four feet and then separated into two pieces. From the V-shaped gap, I was able to take my shots at the guards while safely hiding behind its massive trunk. It was time to get started.

If possible, I hoped to pick off two guards from each location. Taking careful aim at the guard to the right of the one directly in front of me, I held my breath, and fired. Bull's eye! I hit the guard on his right shoulder. He slapped at his shoulder as if he were killing a biting insect. His action caught the attention of the guard next to him. He soon began to stagger a little, and moments later, dropped to the ground. The guard that was in front of me went over to his fallen comrade, as did the guard on the far right. Looking stunned at each other, the guards looked around to see what had happened. During this time, I took advantage of their confusion and set my sights on the guard that was in front of me. Still unseen by the guards, I fired again, hitting my target in the chest. Slapping himself where the projectile hit him, the guard yelled out, alerting his four remaining friends to take cover before he passed out. During the commotion, I took off for the boulder left of the tree. As I ran, one of the guards spotted me and fired a shot in my direction. As I neared the boulder, I slid in on my stomach like a baseball player stealing second base.

Rising into a crouching position, I kept behind the pudgy-looking boulder, which stood about three feet tall and four feet wide. Looking around the boulder to see where the guards were, I was greeted by a couple of shots that hit the boulder at head level. Lying as flat as I could on my stomach, I looked out from the left side of the boulder. I was just able to see the guard posted on the far left clearly in my scope. Taking careful but quick aim, I fired at my adversary, hitting him in the neck. Ouch! I felt bad about that one, but the lowest part of the guy I could see was his upper shoulders.

After he went down, the guards moved about more nervously. I noticed that they never strayed far from their posted positions, which I found to be very odd. At almost any time, they could have surrounded me. They also never made any attempts to call for reinforcements. They didn't seem to be carrying any means of communication with them.

The skirmish continued for a while from behind the boulder, but eventually, I was able to pick off the guard that stood to the left of the one that was directly in front of me.

I avoided a number of shots taken at me as I made the fifteen-yard dash back to the tree, and then the twenty-five yards to the partial wall. The remaining guards could only be seen from the wall. The guy directly behind the building would be my greatest challenge. I still found it very strange that the guards didn't run for better cover. Instead, they stayed near their posted position and used the building as their protection when possible.

The guard on the far right looked like he needed to change his pants. He was dancing around, trying not to make himself an easy target. I was convinced now that the guards were not expected to leave their posted position, apparently for any reason. I could only imagine what threats they must have received that would keep them obedient under their present circumstances. The dancing guard fired at my position rather frantically, trying to keep me concealed behind the wall. The guy behind the structure aided him while keeping himself hidden. Like before, I got down on my stomach, hoping I would get a clear shot at the dancing guard. It wasn't easy, but during a moment when he slowed down to rest a bit, I shot him in his left shoulder, helping him get the rest he wanted.

One guard left! He wasn't going to make my job easy. When he exposed himself to take a shot at me, he was only visible for a couple of seconds. From my current position on the left side of the wall, there was no way I would get this guard unless he got careless. Moving over to the right side, I had to look through a rather dense gathering of dried roots and vines. Aiming from here, I was barely able to get a sight of his right leg. The denseness of the dead vegetation made moving the gun's barrel around for aiming purposes very difficult.

At one point, I took a shot at his leg but ended up hitting the building instead. This startled the guard to move behind the structure even more. Now, I could see nothing of him at all. Using logic, I knew that if I couldn't see him, he couldn't see me either. I, therefore, took the opportunity to come out from behind the wall and head directly for the illuminated structure while the guard concealed himself. I hid myself at the base of the building, opposite of his position. He still occasionally poked out to fire a shot at the wall where I was. As he did so, I crept along the left side of the structure's base, knowing he would need to be on the right side to hit the wall. I tried to move as quickly and quietly as I could to get myself into position. Once there, I waited for the final guard to take another shot at the wall. When he did, I jumped out and aimed at the guard. When he turned around, I shot him in his chest. Looking completely astonished with his eyes opened wide, he staggered for a few seconds, and fell to the ground.

With all the guards now in La La Land for the next six hours, I made my way into the building. It was quite bright inside due to the excessive illumination, as was indicated from the outside. The room was round and made of huge stones. The entire wall was yellow in color, which only added to the room's brilliance. It was supported by three massive beams, also made of stone. The beams were spaced one hundred twenty degrees apart from each other. Taking up a good portion of the floor was a triangular stage that stood two feet tall. Each corner of the stage was attached to a beam. Within the room lay a large number of objects strewn about the floor, some on the stage, others at ground level. There were also tables that contained objects as well. I rummaged through some of the stuff but saw nothing that appeared to have any rhyme or reason to it, except for one object. It was on a table within a box, a beautiful green glass sphere. It was about an inch or two in diameter. Other than the sphere, all of the objects seemed rather ordinary. The room seemed to be a large assortment of loose material. I almost felt that I was in a huge storage closet.

Against the wall on the left side of the room was a staircase made of stone, which led to a trap door. Finding it unlocked, I slowly opened the door and looked around before fully entering the room. From what I could see, it had the same design as the first floor, except the wall had no color. Entering the room a bit more, I noticed subtle pieces of evidence that the room had been occupied not too long ago. Once I was off the stairs, I closed the trap door behind me. I took a few steps into the room and found myself diving for cover. Apparently, I wasn't alone. Thankfully, the other occupant within the room was a lousy shot, or I'd be dead. The objects on the floor, the stage area, and the beams all made good hiding places. At first, I had no idea where the gunman was. Trying to remain still, I listened for a little while. I ended up hearing a small shuffling toward the back left side of the room.

Arming myself, I scanned the area just mentioned. At first, I saw nothing at all. I tried to move from place to place with as much stealth as I could. Finally, while lying on the floor alongside the stage area, I caught sight of my opponent. I felt horrible lining people up in my sights, but what else could I do? Taking aim, I took the shot, hitting the individual on his left side. It didn't take long before the guy dropped. I felt it safe to assume that the gunman was alone. If not, the skirmish would have revealed any others present. That made a total of seven people I've put to sleep. I felt there would be more, that is, if I didn't die first. It seemed that I would need to be on very high alert for the remainder of my stay. I really must have a good talk with Rich, Nirton, or whatever his name is. I'm getting more out of this adventure than I signed up for.

Looking through the items that lay about, I saw nothing of interest, not even a colorful glass sphere. Heading to the stairs, I reached the trapdoor and pushed it open ever so slowly. Looking around the room from the tiniest crack, I saw that the wall was colored blue. This room seemed to be like the one on the first floor. I continued to creep slowly into the room, trying to keep myself ready to fight if need be. Thankfully, there was no need. Feeling more comfortable, I looked through the items that lay about, finding another colorful glass sphere. This one was located on the stage under a small amount of clutter. The sphere was red this time. It looked like I was going to be making a collection of colorful spheres. I, therefore, returned, very carefully, to the first floor to retrieve the first sphere. I placed it in my backpack and headed for the third floor again. Before leaving, I took notice of the wall color again. When I returned, I placed the red sphere in my pack as well and took note of the wall color. Finding nothing more, I slowly walked to the stone stairs.

I had a good feeling that I was caught up in a pattern. So far, the rooms on the odd-numbered floors had colors to remember, namely, the walls and the spheres. The even-numbered floor had someone waiting to blow my head off. If I'm right, I can expect more action on floor number four above me. Wasting no time, I began to open the trapdoor leading to the fourth floor. Without going into any major detail, it sufficeth me to say that I was right. This time, though, there were two Allorayans to put to sleep. One of them almost put me to sleep permanently. But as I said earlier, these guys have lousy aim, for which I was very thankful.

The same pattern continued for floors five through nine. I found the spheres I was supposed to find on floors five, seven, and nine. The colors of the spheres were yellow, orange, and blue, respectively. As for the walls, their colors were red, green, and orange respectively.

As for my friends on floors six and eight, they proved to be quite menacing, and their aim improved slightly. On floor six, I faced three opponents, as I figured I would. On floor eight, I had to deal with four. Before entering floor six, I picked up a metal object that I used as a shield. It helped protect me quite a bit. I will say that I did a lot of diving, sliding, and throwing objects to take out my opponents. On floor eight, one of them nicked me in the leg. I managed to stop the bleeding once I put the final Allorayan to sleep. It just happened to be the same guy that got me. I shot him in the head, on purpose. He's going to have a double headache.

I was not looking forward to floor number ten. If the pattern stayed the same, there would be five Allorayans to deal with. Be that as it may, I began to climb the stairs to the trapdoor. When I pushed on the door, it did not budge. Then, I noticed five colored squares off to my left, each one with an indentation at its middle. If I wasn't mistaken, they were the same colors as those found on the walls of the odd-numbered floors. The indentations were just the right size for the glass spheres. Looking over notes I had jotted down, I saw that the colored squares were not in the order they appeared on each odd-numbered floor. I stood there pondering why the order of the colors didn't match while reaching out to touch the first square.

Upon contact, the first square's color changed from green to blue. Knowing that the first-floor wall color was yellow, I continued to touch the square until it turned yellow. Moving on to the next square, I changed it to blue. I continued until all five squares were in the correct order. I then took out the green glass sphere I found on floor number one and placed it inside the indentation of square one. In square two, I placed the red sphere. I continued until all the spheres were in their correct squares. When the last sphere went into place, I heard a clicking sound near the trapdoor's handle. I just hope my adversaries didn't hear it. Hoping that the door was now unlocked, I slowly pushed the door open a tiny amount.

Looking around, I saw that the room was the same as all the others before it, except for a holding cell under a force field on top of the stage. Inside the cell, I saw Rich, or Nirton, I mean. He was talking with another person who I assumed to be the one from the battle footage I watched within the pyramid building. I hoped to learn the identity of Nirton's mysterious guest.

The two men were talking excitedly, having spotted me crack the trapdoor open. They discretely and silently tried to send me messages by pointing around the room. I took it that they were pointing out the locations of the guards, although I wasn't totally certain. By now, I had enough of playing with these guys. I went into my gear and pulled out the sonic grenade from Nirton's ship. I also pulled out the earplugs so that I would be able to function after the blast. To be decent, I showed the grenade to Nirton and his companion. They looked at each other for a few moments, then looked back at me again and nodded their approval.

On a signal from Nirton, I advanced my way into the room and tossed the grenade. Nirton and his friend dropped to the ground, covering their ears tightly. Unfortunately, one of the guards caught sight of the odd behavior of Nirton and his companion moments after they fell to the floor. I was then spotted almost immediately. Having myself ready, I fired at the closest guard to me just before he and one of his comrades started shooting at me. The guard I shot went down as the other took cover, alerting the other three as he did. I dropped to the floor and scurried about, trying not to get spotted. Moments later, the grenade went off, blowing things around the room due to the sonic vibrations. When the blast subsided, the room was in shambles. The remaining guards were out for the count. Nonetheless, I went over to each one and shot them in the arm. I needed to make sure that they would stay asleep.

As I went about the room, tranquilizing the guards, I noticed Nirton and his friend stirring on the cell floor. Apparently, the blast still had quite an effect on them, but they soon recovered. Still a bit disoriented, Nirton pointed across the room to a green-illuminated panel on the wall. He then spoke the following as one who was just waking from a deep sleep:

"Lee, touch the panel. It will deactivate the force field around the cell."

Moving across the room, I touched the panel, and a small section of the ceiling lowered itself. Inside the area of the ceiling was a sphere so brilliantly lit that it was painful to look at it. Nirton then continued with a simple command in a slightly sturdier voice:

" You must destroy it."

Pulling out the laser rifle from Nirton's ship, I shot at the sphere. It took three hits before it blew apart. Once destroyed, the lights in the building shut off, and the force field around the cell was deactivated. I then heard Nirton say,

"That ought to keep them inert."

I then thought about the message I got from Nirton that said, "Overcome them with darkness." I now understood that the Allorayans had become heliophiles. Their race, over time, became dependent upon bright environments to give them energy. Without the bright interior of the building, the Allorayans would remain dormant.

Turning on my flashlight, I headed over to Nirton and his companion. I then said,

"Nirton, I presume? It is wonderful to see you again, my friend."

"You are a very clever man, Captain Travis," responded Nirton. "You are quite correct. I am Nirton Zesaq. Please allow me to introduce Caynos Quosh, my colleague and long-time friend."

"Mr. Quosh, how do you do?"

"It is a pleasure to finally meet you, Captain Travis," responded Caynos. "I've heard quite a bit about you."

"Please, call me Lee." I then continued,

"So, Nirton, I saw another name listed on the identification table within the domed central complex building. I recognized it to be the name you spoke about when you initially invited me here."

"Yes, Lee! The name was Axnard Jirroch. He was the President of our city."

"Was," I said.

"Yes, Lee," Nirton responded. "Sadly, President Jirroch was killed during the attack on the city."

"Oh! Nirton, I'm truly sorry. Please…"

"It's alright, Lee. I'm just sorry that you didn't get a chance to meet each other. I know that the two of you would have gotten along very nicely."

"Well, let us depart from here right away. I must retrieve my transport device."

"Your what?" I asked.

"Transport device," Nirton responded. "When we were brought here, the Allorayans took our belongings before locking us up. Gratefully, they didn't examine them too closely."

Nirton then continued, “Lee, may I have a little light over here?”

“Sure, just a moment,” I responded.”

“Ah!” exclaimed Nirton. “There it is. Thank you, Lee. Now we can go. Let’s stand close together.”

“Hold on just a second! I need to know a few things before we go.”

“I understand,” said Nirton, “but we will be much safer if we have our conversation back on Marathrax. I will answer all your questions then. Let us go, my friends.”

Nirton then activated his transport device. It dematerialized the three of us out of the building and re-materialized us onto Nirton’s ship. Once aboard, Nirton took his seat at the command station, cloaked the vessel, and lifted off for Marathrax. I then said to Nirton,

“We’re also going to have to talk about all the cool gadgets and weapons that don’t seem to be anywhere on my ship.”

Nirton and Caynos smiled and laughed.

CHAPTER 33
REUNION

While traveling to Marathrax, I continued the conversation.

"You may be unaware, but when I arrived at Marathrax, I found the city fairly damaged and completely devoid of people."

Nirton and Caynos quickly turned to look at me in astonishment. They spoke no words. I then continued,

"As we approached the planet, I saw four ships leaving Marathrax, one of which left its group and headed over to me. The occupant of the craft started a fight that didn't end well for him. I felt horrible for having taken someone's life, but it was either him or me. Moments later, SAROS picked up high levels of time distortions emanating from the planet. I was then informed that all life readings and distortions were gone. In a state of total disbelief, shock, and depression, I had SAROS continue to monitor the planet while taking me to the third planet so I could get a little rest and think things over.

The planet was so beautiful; I ended up falling asleep for many hours. Feeling better when I awoke, I went to Marathrax and found that SAROS' readings were accurate; the planet was devoid of people. I located your ship and went inside. I found the message you left me lying on your command chair. Having received your instructions to find the five hidden keys, I set out to complete the task."

"After your capture, the domed complex was damaged pretty well. As I went about searching for the keys, I came across an individual hiding in the cylindrical building. He told me his name was Noks. I asked him where all the inhabitants of the city went, and how he managed to avoid whatever happened. He said that he was unsure how he remained unaffected by the incident. He then informed me that the attack on the city damaged a time-altering device that resides under the domed section of the central complex. He explained that the damage to the device knocked the people of the city out of phase with normal time, causing them to disappear, if you will." "Since that time, Noks helped me to look for the keys to get into the room to restore time, or so I thought… It turned out that Noks was an Allorayan spy sent to steal the keys to prevent the city from ever being restored. He confronted me just as I was coming to look for you. He held his weapon on me, demanding the keys I had found. He was disappointed that I only had four keys but wanted them anyway. He didn't want to take the chance that I'd find all five. To make a long story short, I kicked his butt, sending him to the ground. I tied him up, took his weapon, and carried him aboard this ship."

"It was then that SAROS saved the day by informing me that my so-called colleague had come on board to install a new device. SAROS was instructed to ask me the purpose of the instrument, and not to scan it. When I checked out the device after being told its location by SAROS, I saw that it was a bomb set to explode after the ship was engaged. I removed the bomb and threw it as far away from the city and ship as I could. When the crisis was averted, I had SAROS set course for Allorayus. When I got there, I dumped Nok's unconscious mess on the ground and left him there. I then literally fought my way to you."

When I finished my report to Nirton and Caynos, the room remained silent for a while as we all looked at each other. Finally, Nirton spoke,

"Lee, I don't know what to say. I'm quite overwhelmed at the moment. Please, know that I'm very sorry… Did I hear you say that the people are out of phase?"

"Yes, I do believe so, Nirton. In order to restore time to normal, we need all five keys. My trip to find you became a dual purpose."

Nirton then said, "I'm so sorry, my friend. I have put your life in such incredible danger. I assure you that I—"

"Nirton, I know you would never have sent me into harm's path. You could never have possibly known all of the events that would take place. Do not feel bad for another moment longer. We have work to do."

Caynos then spoke, "Nirton, when you described this man to me, you didn't give him nearly the admiration he deserves." He continued, "Captain Travis, you are truly a man of honor. You have risked your life while trying to help a race of people you don't even know. You will always be revered by me as a true hero. I wish to express my sincere gratitude to you. You are welcome in our world anytime."

"Thank you Caynos! I do not feel that I am deserving of such high praise. I just did what needed to be done. That's all!"

Caynos then said, "I shall be sure to add humility to your list of qualities."

"Caynos is right!, said Nirton. You truly are a hero in every sense of the word. What you have done will be honored and remembered by the people of this world always. I, too, sincerely thank you for everything you have endured. You have made the restoration of this city a possibility. Your actions will not go unrewarded."

I then said, "Now it is I who does not know what to say. I truly appreciate your very kind words."

"You are very welcome, my friend," spoke Nirton and Caynos together.

Nirton continued, "I would like to say that you are quite intelligent, Captain Travis. How did you come to know that I was Nirton?"

I replied, "I began putting it together when I tried to get into the domed complex in the center of the city. A computer asked me how many keys I had. I input four keys due to the number I had in my possession. I found myself running at top speed for the door when an alarm sounded. I then hid myself a small distance from the building for a little while. After calming down, I thought about it and returned to input five keys, knowing that there were five keys in total. I remembered that you were still in possession of the fifth key according to the note you left me."

"I am glad that you eventually input five keys," said Nirton. "The computer controls a rather intimidating alarm designed to stop any illegal entries. I'm sorry you ended up finding that out the hard way. The security system totally escaped my mind. I would not have purposely scared you so."

"Forget about it," I said. "Anyway, when I input five keys, the computer offered me three names. I supposed that they were the three individuals who owned the keys of that particular rank. I was confused, however, when I didn't see the name of Richard Gaber on the screen. Then again, why would a man from Earth have a top security-ranking job in another world? Therefore, I initially assumed that you were borrowing the keys from one of the individuals whose names were listed. Not knowing which name to choose, I simply picked one. After guessing wrong and running again, I tried a second time. On my next attempt, I chose the middle name, Nirton Zesaq, and hoped for the best. I was grateful to see that the computer didn't chase me off again.

"Again, I am so sorry, my friend. How greatly I must have frustrated you due to my absent-mindedness. I should have informed you of whose keys you were finding. There was just so much going on, but that is no excuse. I do apologize."

"Like I said before, Nirton, forget it. Everything turned out alright. No harm done——. Anyway, having correctly chosen Nirton Zesaq, it either meant that you were borrowing Nirton's keys, or that you were in fact, Nirton Zesaq. I then remembered the message you left me on your recorder device, which you left on the command chair of your ship. You mentioned that you were in the domed complex and that you removed all five keys for safety. You also mentioned that you had hidden the keys and that you needed me to find them. I supposed that you still could have borrowed them, but it seemed more logical that the keys belonged to you initially."

“The last piece of evidence I had that confirmed my suspicions was the ‘x’ on your homing device, which gave the location of Nirton Zesaq on Allorayus.

Well, I need to confess: I pestered SAROS with a few questions dealing with my suspicions. Eventually, SAROS let the cat out of the bag. She let me know that you and Richard Gaber were one and the same. All along, I assumed that I was looking for Richard Gaber, who had been captured in the city’s initial attack. I see now that I was in error.”

“It’s mind-boggling to think that you lived on Earth for as long as you did, and nobody ever knew your true identity. At least it explains why you believed in the company’s goal of achieving light travel so much, having already achieved the technology. Even the length of time you took to announce your discovery to me was amazing. Well done!”

Nirton then spoke, “Lee, I hope that you are not upset with me or my people for handling our first contact the way we did. We had much to learn about your race. You and your people are very interesting. We see that we can look forward to an ongoing friendship between our worlds, which I hope will come to pass.”

“Nirton, I’m sure your friendship will be more than welcome. It may take a small amount of time, seeing our race is still maturing, but I’m sure it will happen. I look forward to it.”

“As do I, my friend—. Come now, we have a city to restore.”

“That reminds me,” I said. “You wrote in one of your notes that you have the key that is missing. Do you have it with you now?”

Nirton replied, “Look!”

I looked closely at Nirton’s belt buckle and saw a rectangular object with a hole in its middle.

“You’re kidding,” I said.

Nirton then responded, “Because of its shape, it did not draw attention to itself. I trust you have the other four keys.”

“Yes, they are here in my backpack. I almost lost them due to the confrontation with Noks. Apparently, the Allorayans believe that your race means to finish theirs off with your advanced technology.”

“The Allorayans have a lot of paranoid beliefs,” said Nirton. “All attempts to get through to them have failed. All we can do is continue to try.”

After our conversation, Nirton, Caynos, and I completed the remainder of the trip to Marathrax, speaking very few words. When the ship landed at last, we made our way to the domed complex in the hopes of restoring the city to its former glory.

Chapter 34
Restoration

Entering the domed building with people who knew what they were doing was a lot easier. No alarms going off, no running from the building; I tell you, it was wonderful. With all five keys, I was finally going to see the central room under the dome. It would only be a few more moments as Nirton installed all the keys. Watching the grand doors open demanded one's attention. Nirton and Caynos then turned and invited me to enter. Once inside the room, I felt like I had entered into the heart of a sentient being. Nirton walked up to the main device, which controlled the time-altering function. He took the keys over to an area of the computer that had five featureless indentations within it. He began to move as if to begin placing the keys, and then he stopped. He slowly turned to look at me and spoke the following:

"You should be the one doing this. You have earned it. I can't thank you enough for all you have done."

Nirton then handed me the five keys. I felt dazed. I didn't know what to say. This was the moment I had worked for, and yet, I felt so inadequate as I took the keys from Nirton.

"Nirton, are you sure about this?" I asked.

"I've never been more certain about anything, my friend."

"I'm afraid I don't know the order to put them in. The indentations are featureless. Really, Nirton, you should be the one doing this."

"Just place them in the order of Varyn's name. Do you recall how to do that?"

I looked at Nirton with a puzzled expression. Then, it dawned on me how I needed to proceed. That was what the stone tablets in the buildings were for. I went through the notes that I had been collecting throughout my adventure. I found that the first letter I encountered came from the pyramid structure. The letter on the tablet was Y. Putting the keys in their proper order to spell 'Varyn' meant starting with the spherical key. I looked at Nirton, smiled, and said, "I remember, my friend."

I then placed the spherical key in first, followed by the rectangular key with the hole at its center. The third key was the spiral. The fourth was the pyramid key, and finally, the cubicle key finished the sequence. After inserting all five keys, SAROS said,

"Please indicate the function needed to be completed."

I looked over at Nirton and Caynos, unsure of how to respond. Nirton then spoke,

“SAROS, this is Nirton Zesaq, acting president of the city of Iarlabac. The flow of time being experienced within the city is out of phase with actual time due to the Allorayan attack, which damaged this device. The function to be performed is to synchronize the city’s present time with actual time. Make the necessary preparations, and indicate when ready.”

The lights on the computer began flashing in a random order for a few moments. When finished, SAROS said, “Synchronization process completed; awaiting final command.”

“Proceed with the synchronization process”, Nirton responded.

The computer started giving off a low-frequency pitch, which vibrated the room. I found that I had to cover my ears to block out the sound. It made my body quake rather uncomfortably. About eight to ten seconds later, there was a very bright flash of light. Having been unprepared, I found myself momentarily blinded. As I began regaining my sight by degrees, I saw people slowly fading back into view as time synchronized itself.

I couldn’t believe it! This was unreal and fantastic. I turned to Nirton and Caynos.

“It worked!”, I said. “You’ve done it! The people are back.”

“No, my friend, *you* have done it,” Nirton said. “None of this would have happened if it were not for you. This city owes you a large debt of gratitude. But first, there is something you still need to accomplish.”

“What would that be?” I asked.

“Look at your timepiece. Don’t you need to be somewhere?” Nirton responded.

My eyes widened as I pulled out the timepiece.

“VARYN!” I exclaimed.

The time on the piece read “79:98:68 rotations.”

“I need to be on Varyn by 81:75 rotations,” I said. “Sorry to succeed and run, but I’ve got a long-awaited appointment to keep. See you both when I get back.”

As I left the central room, the faces of people I longed to see were staring at me. It made me feel a little uncomfortable, but I smiled and waved as I passed them. The people looked at me with looks of surprise and interest. Some of them waved back, probably unsure of what they were doing. No words were spoken as I made my way to Nirton’s ship. I felt the gazes of the people burrowing into the back of my head as the door to the craft closed behind me. I looked forward to the time when I would be able to meet them properly. For now, I had a date with

Varyn. Looking through my notes once more, I looked up Varyn's coordinates and input the numbers into the ship's computer.

"Welcome back, Captain!", SAROS greeted.

"Thank you, SAROS. It's great to be back. I believe it is time for a trip to Varyn. Engage!"

"Right away, Captain. The time has finally come after so long a wait."

PART 3

THE VAULTS OF FRIENDSHIP, EXCHANGED

CHAPTER 35
VARYN

"Caynos, I forgot to warn Captain Travis about the Zyndok. Quickly, let's see if we can catch him before he leaves."

"Excuse us, please! Coming through! Excuse us!"

"No!" Nirton exclaimed. "We are too late. He has already lifted off." He then continued, "Good luck, my friend. Remember who you are."

"I believe he will be just fine," responded Caynos. "Captain Travis is a good man."

"Yes, I'm sure you're right," Nirton replied.

As the journey to Varyn continued, SAROS gave the usual information about the approaching destination. It seemed like old times, but it felt like ages had passed since the last time SAROS gave me planetary information. There was something very comforting about it. SAROS then said, "Captain, we will land on Varyn in forty-eight seconds. Varyn is the only natural satellite of Marathrax. It is named after the founder of the inhabited cities of Marathrax. The surface gravity of the satellite measures 64% that of Marathrax. The average surface temperature is -93 degrees Fahrenheit. The atmospheric composition: None."

As SAROS finished her report, the Lotania touched down on the dry and barren surface. From what I could see of it, Varyn had many characteristics like that of Earth's satellite. After hearing about the outside conditions, I automatically retrieved my normal space suit and boots for my excursion. It should have seemed strange not having to dress up in a while, but dressing felt like second nature to me. While I was finishing up, my timepiece started beeping. Looking at it, I saw that it read "81:70:07 rotations." I had very little time left.

Moving at a quicker pace, I exited the ship and walked onto the surface of Varyn. Just as depicted in the mural within the cylindrical building, the mineral-encrusted outcrop of rock stood before me. Not far away was the rock face that would receive the outcrop's beam of light. *This ought to be great,* I thought. *I'm glad I didn't arrive late.* Looking toward the horizon, I could see where Uganthos would make its appearance. It would not be long now. The timepiece now read "81:73:12 rotations."

Taking a quick look around me, I became stunned as my eyes fixed themselves on the most brilliant constellation I had ever seen. I couldn't believe how I missed it when exiting the craft. I recognized it to be the star pattern of Varyn from the spherical building. The stars appeared to display many colors, with the chest star being a very bright white. It was difficult to look away from the captivating spectacle, but I needed to check the time again. The clock now read "81:74:82 rotations." What I had come to see was just seconds away. Holding my breath as the final seconds slipped into the past, 81:75:00 rotations occurred.

Right on time, Uganthos peeked above the horizon. It was incredibly beautiful to behold. Its intense orange gas could almost be seen to swirl as it rose higher above the horizon with every passing second. On the ground, I saw Uganthos' light creep slowly toward the outcrop of rock. The information I read before stated that the beam of light would start at precisely 81:99:00 rotations. I watched with unbroken attention as the light began to climb the outcrop.

When the light reached the encrusted minerals, I stared with my eyes opened wider than I thought I could ever open them. Just then, a beam of light projected from the rock as if coming from within the outcrop itself. I was so excited; I caught myself practically jumping up and down. The beam shot over to the rock face, just as seen in the mural. Knowing that I only had until 82:00:00 rotations before the beam ended, I hurried the best I could over to the rock face. Not long after I reached the wall of rock, the beam stopped, leaving behind a bright, circular spot.

It was as if the rock had been superheated by the beam of light, except for the fact that the rock face was not molten where the beam made contact. From what I read in the pamphlet in the spiral building, I was supposed to touch the glowing spot before it faded away. What would happen then was left as a surprise. I just remembered the information saying that there would be new doors to explore. Still excited and breathing a bit heavier than normal, I reached out and touched the glowing circle of light. When I did, a small section of the rock face opened up, revealing an entrance to an inner chamber. I guess they weren't kidding when they said new doors. I could tell that my adventure on Varyn was just beginning.

CHAPTER 36
WITHIN THE ROCK FACE: MUSICAL TONES

Incredible! A concealed cave! I stood there staring at the entrance for a few moments. Having no idea how long the opening would remain, I stepped inside the chamber. I could only imagine what secrets existed within, but I would not need to imagine for much longer. After I entered, I saw a device in front of me. My attention was taken away from it momentarily as I heard the door to the cave close behind me. I don't know why, but seeing the door close gave me an unsettling feeling. I guess I felt a little claustrophobic for a moment. Thankfully, the feeling passed just as quickly as it came.

I couldn't tell what, but after the door closed, something about the room changed. I then noticed that the device in front of the door had something displayed on it that wasn't there before. It was a message written in Dhirniy that read: "Atmosphere, temperature, and pressure stabilized. You may now remove your helmet."

I didn't know what to make of that. I just knew that I was a bit reluctant to remove my headgear. I had no reason to mistrust the Dhirniy people, but any number of malfunctions could have occurred. Who knows when the last time this room was occupied? Of course, I did say that I felt a change in the room after the door closed. Thinking it over, I began to remove my helmet ever so slowly. At the first sign of discomfort, sickness, or impending death, I would immediately put my helmet on again.

As you may expect, I was being overly paranoid. Whatever device was controlling the conditions of the room was obviously in excellent working order. It was nice to remove my helmet once again. Now that I had the chance to look around in greater detail, I saw that the room didn't have much in it. In the center of the room, there stood a pillar of rock about three feet in height. Atop the pillar was a small instrument in the shape of a harp. It was finely crafted with twelve strings that seemed to be vibrating.

Looking up, I saw that the walls were white, except for the left wall from where I entered the room. This wall had a colorful design on it, constructed of geometric shapes.

"Not again!" I said aloud.

Taking a few moments, I drew the colorful pattern into my notebook, knowing that it would be important at some time in the future. Finishing my observation, the only thing left that I could see in the room was an outline of a door on the right wall. To the right of the outline was a rectangular unlit panel. Next to the unlit panel was another outline that was small and square.

While checking out the outlines on the wall, I thought I heard something. I turned around quickly to discover whatever I thought I may have heard. The room seemed as it did a few moments ago. I could see nothing out of the ordinary. As the seconds passed, the sound grew. Soon, the sound became loud enough that I recognized it to be musical notes being played. I quickly looked over at the harp-like instrument in the center of the room. Earlier, the strings seemed to be vibrating; I was now able to see that I was right. The strings were now so agitated that the vibrations nearly caused the instrument to vibrate off the pillar. I ran over to it just moments before it fell off the pillar and into my hands. Looking at the instrument up close, I saw that it was much more beautiful than I gave it credit. It then dawned on me that the musical notes were no longer playing. They stopped when the instrument fell off the pillar. To see if I was right, I placed the instrument back on top of the pillar.

Within moments, the musical notes became audible again. After listening for a while, I recognized that the notes were repeating themselves in a cycle. It was a string of ten notes that would play over and over again. I wasn't quite sure exactly what was going on, but a good guess was that I was expected to play the ten notes in the same order I was hearing them played. I tried to play the notes, but the only thing that happened was the continual repetition of the cycle. It looked like I needed to stop the instrument from playing so that *I* could play it. This would, of course, mean that I would not hear the notes being played anymore. Needless to say, this took me a while to accomplish because I had to listen, play, re-listen, and replay several times before I got it right.

Eventually, I played the correct sequence of notes. I then heard a clicking sound behind me. I turned to see the panel on the wall lit up, and the small square outline opened. Inside the opening was a beautiful key in the shape of a rhombus. This key was even more beautiful than the keys I recently collected on Marathrax. It was five inches long and golden in color. The workmanship was indescribable. No doubt, this key was worth a fortune. It seemed that I would be hunting for keys again.

Moving on, I touched the lit rectangle on the wall and the door outline opened up. Looking inside, I could see a flight of stairs that led downward to another door outline. To the right of the door, where the lit panel should be, there was an indentation, presumably for the rhombus key. While walking down the stairs to the other door outline, the door to the music room closed behind me. I turned quickly, feeling a moment of fright. I didn't expect the door to close, although I should have. I'm glad that I saw the rhombus-shaped indentation on the wall before entering the corridor. I would not wish to get trapped in here because I forgot the key. I say this because I saw no way to re-open the door to the music room. Forgetting the key would be a death sentence unless found by someone, and that's only if they knew you were missing.

After inserting the rhombus key into the wall, as expected, the door outline opened, showing another room beyond. I started to make my way into the new room when I remembered to take the rhombus key out of the wall. I needed to remember to take the keys everywhere I went. Having the key stored away, I entered the new chamber.

CHAPTER 37
THE FOUNDER'S ROOM

Once inside, the door closed behind me, unsettling me again. Looking at a door that had no apparent way of opening was a bit unnerving. Trying not to focus on my feelings of being entombed, I looked around the room to see what challenge it had for me. In front of me was my next exit. The room was designed the same as the last one with the door and square outlines and an unlit rectangle.

To my left was an odd sight. It looked like a deliberate set of five stone steps. I'm sure I was right because at the top step was a porthole of sorts, enabling me to look outside. The hole was covered with a thick, clear piece of glass to keep Varyn's natural conditions from coming in. I was quite grateful for this, seeing I wasn't wearing my helmet. When I stepped on the top stair, the lights in the room dimmed to almost nothing. This enabled me to see the outside without a competing glare. I was able to see that the porthole was just above ground level. From my new perspective, I was able to see the constellations quite easily. I also noticed how much Uganthos had risen since striking the outcrop of rock.

Coming down the natural steps, I headed over to what I was sure to be the room's major attraction, as the lights returned to their original brightness. Embedded in the left wall were seventeen clear, circular glass coverings. The pieces of glass were connected with lines that made up the figure of a man. The figure was none other than Varyn, I presumed. I remembered having seen this pattern a few times before. Examining the clear pieces of glass, I touched one of them, which made it turn white. Touching it a second time changed its appearance to red. By continuing to touch the glass, it cycled through the colors of the visible spectrum, ending with the original clear appearance. It didn't take a rocket scientist to figure out what needed to be done; I just needed to know the correct colors of Varyn's stars.

When presented with them in the spherical building's planetarium, I did not write them down. The Dhirniy already planned on me forgetting to write down the colors, so I assumed this is why the room had a porthole. Of course, while on Varyn's surface, I didn't take a good look at the sky. I just described it as dark. *YOU THINK? IT'S ONLY OUTER SPACE! Of course, it's going to be dark.* I'm telling you, sometimes I find myself to be an unending source of pure astonishment.

Walking back over to the porthole, the lights dimmed again when I stood on the top step. Looking outside, I searched the DARK sky for Varyn. I was having a hard time getting over those brilliant words of mine. That had to be one of my all-time dumbest things to say. *Ok! Let it go!*

My search for Varyn didn't take long. If you remember, Varyn was the brightest constellation I had ever seen. Examining it as well as I could from my position, I wrote down the colors of the stars making up the Varyn constellation. I should say, all but one star. The outcrop of rock obscured one star due to the star pattern's position in the sky relative to mine. Taking my information over to the pattern on the wall, I recreated the correct color pattern for the Varyn constellation, except for that one star, of course. I played around with the remaining piece of glass and found that the missing color was blue by the process of elimination.

When the final color was introduced to the pattern, I heard a familiar clicking sound off to my right. Looking over, I saw an illuminated rectangular panel to the right of the door outline. I also noticed that the square outline opened as well. Inside the opening, I found an exquisitely beautiful octagonal key. Like the rhombus key, it was golden, five inches long, and probably worth a fortune. After putting the key in my pack, I touched the lit rectangular panel and found myself looking at a staircase that led down to another door outline. I could only wonder how many times I'd be doing this, and how deep into Varyn's surface I would be traveling.

Being sure to have both keys, I stepped into the new corridor and walked down the stairs. Just as before, the door to the previous room closed tightly behind me. This door outline differed only in the fact that it had an indentation on both the right and left sides of the door. The left side had the familiar rhombus-shaped indentation, and the right had the new octagonal-shaped indentation. Placing both keys in their proper spots, the door outline opened to reveal another room. Retrieving my keys, I stepped into the new chamber.

CHAPTER 38
COLORFUL GEOMETRIC PHANTASM

This room was a bit different, which I called the rainbow room. I say this because the walls were a continuous spectrum except for one area. On the left wall, there was a puzzle where the color of the wall was an indigo/violet. Where the puzzle sat, the wall remained neutral. On the right wall was a message written in the Dhirniy language. It was written right over the yellow section of the visible spectrum. After translation, the message read:

"Remember the test and why you are here; be kind and generous, and do not fear."

I can't say that the words left me feeling at ease, but it did no good fretting over them, at least not yet. Like the room before this one, the door outline and its companions were on the wall in front of me. As for the puzzle on the left wall, it looked very similar to the Varyn puzzle in the previous room. It was one of those touch-and-go puzzles where you touch the glass covering, and the color changes; only these glass coverings had a button on top of them. I wished I knew what colors the puzzle required. Looking through my notes from information gathered on Marathrax, I couldn't find anything to help me solve the puzzle. The puzzle in the domed complex was another similar puzzle, but it didn't help me with this one.

Continuing to search through my notes in the event I missed something, I found that I was looking in the wrong place. The answer was not presented from information gathered on Marathrax, but from this very cave. In the first room with the harp, there was a colorful pattern constructed of geometric shapes. The answer to the puzzle in this room was displayed on the wall of the music room. I'm glad I wrote it down; otherwise, I'd be sunk. As I mentioned already, I could find no apparent way to backtrack into previous rooms once a door closed behind me.

As for this puzzle, just looking at the answer practically gave me a headache. It was very complex, to say the least. The puzzle required many geometric shapes. Without counting, I'd guess there were a good twenty-five to thirty different pieces.

Well, the sooner I got started, the faster I'd get done. I could tell I was really tired again. I just didn't feel very enthusiastic about tackling the puzzle at this time. Instead, I just wanted to lay down on the ground and rest. Using my pack as a pillow, I fell asleep in just a few moments.

Just how long I was asleep, I couldn't tell you. I just knew that I needed the rest. After waking up, my mind felt more invigorated. All of a sudden, the puzzle didn't seem as ominous as it did before. Taking another look at the geometric phantasm, I could almost sense it staring back at me, daring me to solve it. I took that as a challenge! I, therefore, touched one of the glass coverings to see if it would change colors as the Varyn puzzle pieces did in the previous room. I was pleased to see that the pieces responded the same, but my pleasure was very short-lived. To start with, the puzzle pieces acted all wrong. After touching the glass covering, the color would show up as expected but would change to a different color on its own a couple of seconds later. After another couple of seconds, it would change again, continuing in this manner. *How rude!*

It took me a few moments to catch on what the buttons were for. Touching the button would stop the color from changing, locking it in place, so to speak. There was just one further problem. The color I locked in was not the color that ended up being displayed. For example, I tried to lock in the color red where the puzzle needed red, but after locking it in, the color blue would be displayed. I tried it more than once, thinking I may have done something wrong, but the end result was the same every time. Checking other colors, I input yellow where the puzzle required it, and I ended up with violet. This puzzle was messed up!

It was bad enough that I had to input as many as thirty colors into this thing, but now, I had to figure out what the true color would end up being. This was a puzzle within a puzzle. *Just what I needed!*

I'm so glad that I chose to take a nap first. I probably would have lost it otherwise. Anyway! I ended up learning that a clear covering would change to red, red would change to blue, orange to yellow, yellow to violet, green to red, blue to orange, violet to green, and finally, white to clear. Once I had the color-changing sequences figured out, solving the puzzle became a lot easier. Even so, it still took me about twenty minutes to complete. When finished, it was very impressive to look at. Being three dimensional, it was a rather hypnotic sight to behold.

When I input the last color, I heard the familiar clicking sound off to my right. I head over to the door outline, where I retrieved a beautiful cylindrical key from the wall. After storing it with the other treasures I recently collected, I touched the lit rectangle, which opened the door. As I walked into the corridor, I braced myself for the unnerving sound of the door closing behind me. I never did get used to that noise. It still troubles me a bit when I think about it even now.

After the door closed, I descended the stairs, moving deeper into the belly of Varyn. At the new door outline, I saw three indentations—the rhombus on the left, the octagonal on the right, and the new cylindrical indentation above the door outline. I placed the three keys in their niches, and a new chamber was presented to me. From where I stood, I could see that this room was very different. I received a feeling of excitement as I gazed into it. I retrieved the keys, still looking at the sight before me. I placed the keys into my pack as I entered the room that would change my life forever.

CHAPTER 39
THE PORTALS TO THE WORLDS

I didn't walk very far before stopping. I could see that the chamber was oval in shape. I stood on a pathway that normally would encircle the entire room. I say normally because, to my immediate right and left, the pathway was incomplete. I saw that the pathway continued on, well out of reach, after a large gap that yawned before me. In front of me was a stone bridge that led into what looked like a maze. I say this because just beyond the bridge, my path split into two, giving me a choice of which way to go. The bridge itself crossed over an abyss of unknown depth, which separated the outer path from the inner maze. It was like a mote encircling a castle; only in this case, it was a mote of death. Trying to jump across the gap to walk along the outer path would have been foolish. As I already mentioned, the gap was way too large.

From where I stood, I was able to see that along the outer path, there were doors in the walls. To the right of each door visible to me, there was a lit rectangle waiting to be touched. I could only hope that penetrating the maze would place me on the outer path. Seeing no other choice, I walked over the bridge toward the entrance to the maze. At this point, I needed to make a decision: do I go right or left? I generally did horribly when presented with a fifty-fifty choice, so I wasn't comfortable at this time. I've always heard that choosing the right path within a maze whenever possible will yield good results. I, therefore, chose to go left; I just felt that going left was the correct move to make. I know I told you about a song called "Choose the Right" earlier, but I still chose to go left. I will probably end up paying for it.

Like any maze, I was offered a number of choices as I wandered aimlessly through it. It didn't seem like it would be terribly difficult from the outside, because it didn't look to be that large an area. Regardless of its appearance, I was lost. For the first ten minutes, I didn't feel like I was getting anywhere. After about fifteen minutes of walking that path of insanity, I came to a post with a lit circular globe at its top. It looked like a miniature lamppost that you might see along a street. Examining it closely, there didn't seem to be anything special about it, but finding an object like this in a maze had to have some significance.

Continuing to look it over, I began feeling for anything odd, seeing that I couldn't find anything with my eyes. I started at its base and worked my way up. I found nothing until I reached the top. It was the globe! When I touched it, I heard something going on in a different area of the maze. I couldn't see what the action was, but it sounded as though stone was scraping against stone.

The sound seemed to come from somewhere behind me. I wasn't sure, but I bet it was the outer pathway at the beginning of the maze connecting to form a complete passage.

Hearing no further sounds, I continued trudging through the maze, every step seeming to be more difficult than the last. A couple of minutes later, I saw what looked to be the maze's exit on my left. It was a metal door that was shut and locked. What a depressing sight.

Dreading the moment, I looked to see if there was any way to get on top of the wall. This would certainly give me a much better view of the maze. Perhaps I could use something from my pack.

As I thought of a way to cheat my way out of the maze, I caught sight of a table that was nearby. On top of the table was a familiar metal box, but this one was much longer. Putting my ideas of cheating aside for a moment, I headed over to the table to check out the box. It had no lock of any kind, so I opened it and saw two items within. One was a message that translated to say:

"Three of eight you have of your keys; move quick, look sharp, and then you must flee."

The part about having three keys, I understood; the rest about moving quickly and fleeing made me a bit nervous. I didn't like the idea of having to move quickly and fleeing. *Flee from what?*

Beginning to talk to myself, I then said aloud,

"What is this other thing in the box?"

I picked up the remaining item. In my hand was a cone-shaped object that opened near the pointed end. The cone measured about two feet in length, which explained the need for a larger box. I'd say that the top six inches of the cone opened up. Inside, I thought I saw what looked like spiral stairs. There were other items, but I couldn't identify what they were. It was almost like looking into a huge doll house with all its miniaturized furniture. I wondered what the cone was supposed to mean, as well as the message.

Pondering as I went on, I came to another lit post about two minutes later. Wasting no time, I touched the lit globe and heard some more noises. This time, it didn't sound like stone on stone, but more like squeaking metal in motion. *That could only be one thing,* I thought. Backtracking to where I saw the maze's exit, I cheerfully beheld that the door was now open. Moving quickly to get out of the maze, I passed through the door and found another stone bridge spanning over the abyss of death. Walking carefully, I crossed over the bridge onto the outer pathway. *I DID IT!* I was now free to investigate the doors.

From my new position, I could see that there were six doors in total. They were located in the two, four, six, eight, ten, and twelve o'clock positions. As I previously mentioned, all of the doors, except the twelve o'clock door, had lit rectangles. I was sure that the six o'clock door didn't have a lit rectangle when I first entered the room, but it did now. The opening of the maze's exit may have activated it for all I knew.

Being nearest to the door in the twelve o'clock position, I looked it over closely. As I mentioned, there was a rectangular panel to the right of the door, but it was not lit. On the door itself, there were eight indentations that acted as keyholes. There was an indentation for all eight keys, only three of which I had at the moment. *Whatever is behind this door, it has to be special,* I thought. This door wasn't going to open for a while.

Making my way around the outer path, I saw that the stone-on-stone sound was the pathway having completed itself. The gaps to the right and left of the maze's entrance were gone. Better late than never, I guess.

I now needed to find the remaining five keys. Behind each door somewhere, I assumed there was a key waiting for me. I wondered what types of rooms I would find behind each door. I hoped that the puzzles wouldn't be too difficult to solve. Well, there was only one way to find out; I needed to choose a door.

Finding myself next to the eight o'clock door after walking along the outer path, I started there. When I touched the lit rectangular panel, the door slid open. Looking through the door, my eyes widened as my jaw practically hit the ground. Reflexively, I jumped back as fast as I could, falling to the ground. Breathing heavily in very short and rapid breaths, I found myself shaking as I stared out into the vastness of space. I felt paralyzed as I sat gazing upon a beautiful scene of a planet surrounded by countless stars. Very carefully rising to my feet, I looked through the door from different angles, staying away from it in the process. From the right side of the door, I thought I saw what appeared to be Uganthos. This was unreal! I was so freaked out I couldn't stop shaking for several minutes.

I finally calmed down and gained enough nerve to approach the door a little bit more. I felt no gravitational pull trying to suck me out into space. I, therefore, took another step closer and practically had to change my pants when I heard a voice begin to talk. The following words were spoken:

"Hello, traveler. My name is Varyn Keb. I welcome you to one of the portals of the worlds. You see before you the planet Tarjil, the closest planet to the Sun. Measuring 4,728.4 Safilons in diameter, it is the second smallest of the planets. Tarjil has an average surface temperature of 282 Laemids, making it the system's hottest planet. Although the surface is devoid of life, I'm sure you will enjoy your visit. While you are there, you will not need to worry about the natural conditions of the planet. You will be protected by an energy field that will surround your entire body. As you move, it will move with you. You only need to step through the portal, and you'll be on your way. Enjoy your visit."

YOU HAVE GOT TO BE KIDDING ME, I thought! I was expected to just walk through a door that led out into space. *I DON'T THINK SO!*

Alright! Hold on! This has obviously been done by others many times. Perhaps this was another achievement of Dhirniy technology. It must be perfectly safe.

I CAN'T BELIEVE I'M SAYING THIS. I MUST BE TOTALLY INSANE!

On the other hand, if I didn't try, I was going to be stuck in this room for who knows how long. It appeared the only way out was through this door.

I could feel my heart pounding so hard it felt like it was trying to escape my chest. I tried to take deep breaths to calm my nerves, but it didn't work. Clenching my fists and holding my breath, I walked right up to the door frame. I was one step away from the vacuum of space. I thought of the infamous words that Neil Armstrong spoke so many years ago. I then closed my eyes and took the final step.

CHAPTER 40
TARJIL

Even though I knew I stepped forward, I felt no further sensation. I cracked my eyes open a little to find myself zeroing in on the planet Tarjil. Approaching a planet from within a ship is one thing; doing it without a ship is indescribable. I could tell that my velocity toward the planet was incredibly fast despite the lack of sensation. Now, I knew what an asteroid feels like as it approaches a body in space.

The closer I got to Tarjil, the more detail I was able to make out on its surface. The planet had no atmosphere to speak of and was truly stark. As I neared the surface, my velocity decreased until I came to a soft and very safe landing. Immediately after touchdown, I whirled myself around to see an illuminated archway. I didn't see it when I landed, but there it was. It stood about seven feet in height and four feet wide. It glowed with the appearance of molten gold, and was truly quite a sight to behold. I greatly hoped that the archway would stick around and be my exit back to Varyn.

Looking around, I saw that the planet existed under a dark sky, seeing that it had no atmosphere. Even so, its close proximity to Uganthos provided plenty of light. I enjoyed being free of my helmet despite the planet's natural conditions, which would have normally ended my life within seconds. I couldn't see the energy field that was protecting me, but I did see a huge cone-shaped structure about two to three hundred yards away. Seeing the cone-shaped building reminded me of the cone object I found in the maze. It was obvious where I needed to go next. The only problem with getting to the structure was the huge gorge separating me from it.

The gorge had to be a minimum of fifty feet in depth, with other areas deeper still. It looked like a stone quarry that was in the process of being excavated. Spanning from my end of the gorge to the other was a network of railroad tracks. A rather aerodynamic-looking vehicle sat on the tracks not far from where I stood. I didn't know the reason for the vehicle design, seeing Tarjil was an airless world, but no matter. Looking at the transport brought back tense memories of a similar ride recently taken on the asteroid Vesta. I watched as the asteroid broke apart in space due to structural damage caused by several Vestaquakes. I barely got out of there alive, and now it needed to be done again.

Reluctantly, I approached the vehicle, which sat perched on the edge of the gorge. It opened like one of those fancy old cars of the past. You know, like mine. Inside the vehicle, the main screen had a message displayed. After translation, it read:

S 1, 2, 5, 7, 9, 10, 12, 13, 14,16, 17, 19, 21, 23, 25, L 4, 8, 11, 20, 22, and R 3, 6, 15, 18, 24.

Looking at the message, I was sure these letters and numbers were the keys to traversing the gorge. The S, L, and R stood for straight, left, and right respectively. The highest number was twenty-five, telling me that I would stay busy while crossing the gorge. Examining the controls, I was happy to see that it had a braking system in case of navigational error. Otherwise, a mistake would most likely send me down about fifty-plus feet to my doom.

The ignition, so to speak, was an illuminated circular button below the main screen. I needed to be ready in every way before touching it. To make the instructions easier to follow, I wrote down the traversing steps in order from one to twenty-five. I took my time familiarizing myself with the controls and how to comfortably reach them in a split second. I didn't know how fast I'd be moving, and therefore, needed to be ready for anything. After saying a prayer, I exhaled deeply and slowly through very puffy cheeks. It was time! I touched the ignition button, and the transport took off. It was a very stressful ride. I tried to focus on the directions, and not the abyss beneath me.

Now, I don't wish to bore you with every detail of my arduous trip, so let it be sufficient to say that I eventually made it over safely. I had to use the brake and back up the vehicle numerous times. I gave up counting how many times after a while. The trip took me about forty-five minutes to complete, and I hoped I would never have to do anything like it again. I knew I was hoping in vain though. I'd have to come back eventually, and the only way to do it was over the gorge.

Although completely exhausted and mentally blown, I exited the transport as quickly as I could before taking a rest. I had no desire for any unexpected trips. I sat upon the barren ground, trying to prepare myself to move on. From where I sat, I could see the majority of the cone structure standing alone in the dark. Its white exterior made it look rather ghostly, being bathed in the reddish glow of Uganthos. It wasn't exactly a welcoming sight, but an item I needed resided inside, and I was going to get it. I had to watch my steps while approaching the cone. I came across a crack in the ground, and almost fell into a deep crater. Despite my near accidents, the walk only took eight minutes, but it seemed ten times longer.

At its base, I looked up at the cone's pinnacle. It was a strange building, having an occasional window here and there. The top half had more windows than the bottom half. Looking back down, I continued checking out my immediate area. I began walking around the cone's perimeter in search of its entrance. When found, I cautiously entered and saw that the room was exceedingly spacious. Looking up almost made me dizzy. I was also reminded of the model cone I found in the maze. I remember opening the model and seeing a spiral staircase. The actual building also had an immense spiral staircase winding up the outer wall. Quite a ways off the ground, I saw a walkway and platform. The walkway began where the staircase ended. I noticed that the stairs were not the only way to the platform. In the center of the room, there was a clear chute that held a turbo lift.

Let's see now, a loooong staircase, or a ride on a turbo lift. Hmmm! That's a tough choice. After serious deliberation, which took a whole 0.64 seconds, I chose the staircase. Just kidding! I head over to the turbo lift.

I can tell you that the lift didn't seem like much to look at. I'd seen elevators that were fancier than this thing, but no matter! A bland turbo lift beats a loooong staircase any day.

The lift had only one illuminated button in it. Perhaps it only goes one way. The important thing, is that it would get me to the platform. If I needed to use the stairs on the way down, that wouldn't be too bad. At least I would have gravity working with me. Touching Alphard, which is Arabic for The Solitary One, the lift shot upward at a rather remarkable speed. The ride was over in about four seconds. I really didn't have any chance to look around while ascending the shaft. It was a bit disappointing, actually.

The door to the lift opened as I was finishing up my whining. About five yards away stood a table with a metal box on top of it. Disembarking the lift, I made my way over to the table. A startling heavy vibration almost had me jump out of my socks. Turning quickly, I saw the lift jerk around in the shaft for a second or two and then fall straight to the ground below. I imagined hearing the lift making a spine-tingling crunch as it blew apart on impact, destroying part of the lower shaft with it. I stood there looking down at the wreckage for a few moments. I guess I was wrong when I said, maybe it only goes one way.

Don't mind me! Near-death experiences trigger sick jokes so as to lighten the moment.

Trying to refocus my attention, I approached the table for a second time. Reaching for the metal box, I opened it to find the spherical key. Like its family members, it was extremely beautiful. The craftsmanship was extraordinary. To keep it safe, I put it in my pack before making my way to the staircase. I looked around once more to make sure I didn't overlook anything. There wasn't really anything to miss, so I began my descent. While walking, I was treated to a view from one of the windows that happened to be close to the stairs. The aerial view was pretty nice and would have been much better if it wasn't such a barren landscape.

Once on the ground, I left the structure, dreading the thought of having to traverse that gorge again. To make things seem worse, all of my maneuvers would now need to be done in reverse.

When I arrived at the gorge, I wrote down my reverse course before entering the transport of stress. Again, I will spare you the mind-numbing details of my trip back. I will say that I almost didn't stop on time after making a mistake. The vehicle came alarming close to one of the dead ends, but I'd rather not ponder on that. Thankfully, making it back alive, I gladly exited the transport and headed back to the glowing archway. I was very happy to see that it stayed put. Hoping that it would now bring me back, I stepped through the opening. After doing so, I was treated to another beautiful trip through space, back to the satellite Varyn. This time, I kept my eyes open for the entire trip. It was incredibly breathtaking. I could get used to this.

When the journey ended, I found myself standing in the maze room once again. Closing the door to the planet Tarjil, as crazy as that sounds, I made my way to the next door in the ten o'clock position. As I reached out to touch the lit rectangle alongside the door, a thought came to me. For the safety of the keys, I should insert them in the twelve o'clock door before heading out again. I would insert them one at a time until I finished.

I walked to the door in the twelve o'clock position and inserted the four keys I had collected so far. Feeling less stressed, I headed back to the ten o'clock door and touched the lit rectangle. When the door opened, I felt myself instinctively lurch back once again. I didn't actually move, but I felt the jolt. Again, I found myself staring with widened eyes into the dizzying expanse before me. To my surprise and gladness, I recognized the planet in front of me this time. It was planet three, the water world. I really loved visiting there before. I was very happy to have the opportunity to return. When I took a step closer to the door, Varyn startled me again. I forgot about his introductory speech. Varyn spoke the following words:

"Hello, traveler! My name is Varyn Keb. I welcome you to one of the portals of the worlds. You see before you the planet Dremida, the third planet from the Sun. Measuring an impressive 17,339.1 Safilons in diameter, it is one of this system's larger planets. Dremida has an average surface temperature of 72 Laemids, making it a very welcoming place to rest. This beautiful planet is quite unique in the fact that its surface is 97% water. It is truly an amazing spectacle to behold. While you are there, you will be protected by an energy field that will surround your entire body. As you move, it will move with you. You only need to step through the portal, and you'll be on your way. Enjoy your visit."

Even though I had done this once already, I still found myself feeling overwhelmed as I stared out into space. I can also say that I felt very excited to spacewalk again, especially knowing I was going back to the water world.

Of course! The water world! I knew the cone structure seemed familiar to me. I saw a pointed structure sticking out of the water when I last visited. And the shimmer I saw in the sky must be a distortion of the portal connecting Varyn to Dremida. Feeling more confident, but not without butterflies in my stomach, I stepped forward into the land of endless night once again.

CHAPTER 41
DREMIDA

Let me tell you, traveling through space without a ship is a major rush. It really got my heart pumping and adrenaline flowing. I never could get enough of it. As I approached Dremida's watery surface, I tried to look back to see if I was right about the atmospheric distortion. A few seconds later, I had my answer. Bull's eye! I entered the area right where I discovered the distortion, which wasn't far from Dremida's only area of land. Looking forward again, I saw the cone structure and the water just a few feet away from me. Instinctively, I took a deep breath as I plunged into the depths. I then remembered that I was being protected by an energy field and started to breathe again. It was very cool being able to breathe underwater. I just kept praying that the energy field wouldn't malfunction. The end result would be quite devastating to me.

A few moments later, I soft landed on the ocean floor about twenty feet from the building's entrance. It took a little effort to get used to seeing objects underwater, but I made the best of it. Behind me was the glowing archway, which really looked interesting as it cast a golden hue into the water. I will never look at a distance of twenty feet the same way again. I couldn't believe how long it took me to enter the structure. I felt exhausted once I got inside.

From what I could tell, the inside of the building was constructed the same as the one on Tarjil. It had the spiral staircase, the turbo lift, and a walkway leading to a metal box sitting on a table. I wasn't overly certain, but it seemed that the metal box was above the surface of the water. It then hit me that Dremida's ocean wasn't very deep in here. Remembering that the cone structures were about sixty feet tall, it made the water's depth about forty feet. I wondered if the depth remained fairly constant all around the globe. That would support one of two major ideas: either the water was placed here on purpose, or it is a young ocean that didn't always exist. I later learned that the latter idea was the correct one.

Dremida was not always a water planet. It was once a world very much like Marathrax, only a bit warmer and more humid. About three and a half million years ago, a catastrophic event took place dealing with Uganthos. A shockwave emanated from the Sun, which affected Dremida's orbit and tilt. The planet was at its perihelion when the shockwave struck. It is believed that Dremida's closer proximity to Uganthos was enough to cause the changes. Over the next three million years, the humidity levels increased, causing it to rain the majority of the time. Somehow, the planet began to regain its original orbital pattern, which stopped the excessive rainfall. Today, it is the paradise I described earlier. The planet Marathrax was not affected due to its orbital position at the time. It was traveling away from Uganthos at a position angle of 148 degrees when the shockwave hit. Sadly, Allorayus was greatly affected, and now finds itself slowly drying up over time. Unfortunately, it has shown no signs of orbital healing like its neighbor. In case you're wondering, I learned all this from Nirton before my adventure ended. Isn't it amazing? All those thoughts came to mind from believing that the walkway and table may be above the water.

Breaking free of my thoughts, I reluctantly made my way over to the turbo lift. After that last joy ride, I'm surprised I moved toward it at all. Stepping into it, I happily saw that the panel to activate it was not lit. I foolishly touched it nonetheless, and thankfully, the lift did not respond. I guess that being underwater for a long period of time put an end to its usefulness, and to its ability to destroy its users.

That left me the choice of climbing the stairs or swimming up. Swimming sounded much better and faster. Anything to avoid the stairs!

Having the protection of an energy field made my swim a more relaxing situation mentally. As for physically, I still felt winded when I reached the top. Trying to keep myself afloat, I couldn't quite reach the walkway from the water. I needed to make my way over to the stairs where the walkway began. Once on the walkway, I took a moment to look out of a nearby window. It was an incredible view, and the smell of the water was intoxicating. I'm very glad the protective shield didn't interfere with my ability to hear and smell on this planet. It would have been a major loss. It was so peaceful here. I just wanted to lie down and sleep again. Don't think that I didn't consider it, but I had a task to complete. I look back at it now and kick myself for not having rested as I wished to do. I mean, what was the rush?

I continued to gaze dreamily for a few moments longer, listening to the sound of the water and the wind. The breeze on my face felt absolutely fantastic.

Finding the need to refocus again, I slowly turned away from the window and made my way over to the metal box on the table. Feeling quite relaxed, I lazily opened the box and saw the rectangular key with a hole in its middle. Just like the others, its beauty went beyond description. I stored the piece of art in my pack, hoping to have it join its family members shortly. Then I thought about how to return to the archway. *Do I swim down or take the stairs?* I normally would avoid the stairs, but the window was near the stairs. It was a tough decision, but I eventually decided to swim down after one last look outside. I must have stood there for another five minutes before entering the water.

Soon, I was on the wet surface of Dremida again. As I began exiting the structure, I froze absolutely still. A few hundred feet away from me, I saw a huge sea creature that looked rather dangerous. Not caring to find out if it was, I planned how I would make it back to the archway without being seen. I was reminded of my recent trip to Europa where I encounter an underwater life form. Refocusing on my present situation, I knew that my chances of moving undetected were practically zero. I needed to reach the archway before the creature reached me. *Could I move twenty feet before the creature moved a few hundred?* I thought. There was only one way to find out. My departure needed to be quick, because once I left the structure, I would be a moving target.

Preparing myself, I began kicking my legs while holding on to the sides of the doorway. I planned on pushing off to give myself a quicker start. Unsure of how fast I could move with a pack on my back, I set off, trying to move as fast and as straight as I could. The thought of the creature getting closer gave me the incentive I needed not to look up or let up. I could feel the muscles in my arms and legs beginning to burn, but I dared not slow down. I knew that I was close to the archway due to the increase of golden light off the sandy ocean floor. Any moment now, I would either pass through the gateway, or be caught by the creature to be made a future meal. Closing my eyes for the last foot or two, I swam hard and fast despite the intense pain I was now enduring. Feeling no tentacles around any part of my body, I opened my eyes. I was relieved and overjoyed to see the beautiful sight of Varyn getting closer every second. Yes! I made it! I later wondered if the creature ever pursued me at all.

After gazing peacefully at the stars as I drifted back, I found myself once again standing within the maze room. It felt wonderful to relax while in space, and I sure could have used some additional R and R time on Dremida again. One day, I hope to return, this time without the sea monster. As for now, though, I had three more keys to retrieve.

As I began to move to the next portal, I felt like an over-saturated sponge due to my recent overexertion. It felt like my legs weighed three times heavier than normal. When I finally reached the twelve o'clock door, I placed my most recent find in its proper location. The door welcomed the key home as it slipped quietly into its keyhole.

Moving on, I slowly shuffled myself over to the door in the two o'clock position. Opening the door, I was treated to another stunning view of the heavens. The planet in the distance was very bright white. The surface features got lost in the natural brightness the planet apparently gave off. Stepping close enough to the door frame activated Varyn, who went on to say the following:

"Hello, my friend! My name is Varyn Keb. I welcome you to one of the portals of the worlds. You see before you the planet Yixelex, the fifth planet from Uganthos. This planet is the system's largest, measuring 27,659.8 Safilons in diameter. The average surface temperature of

Yixelex is a chilly -3 Laemids, making it a great spot for those who love Tros. While on Yixelex, you will be protected by an energy field that will surround your entire body. As you move, it will move with you. Now, you only need to step through the portal, and you'll be on your way. Enjoy your visit!"

"Tros! What is tros?" I said aloud.

I guessed that it might be what snow was called here. I didn't doubt it. I figured that's why the planet was so bright. After squinting at the brilliant planet for a few more moments, I stepped into the void and peacefully headed for Yixelex.

What a name!

CHAPTER 42
YIXELEX

As expected, the trip to Yixelex was very peaceful, except for the brightness of the planet. It almost hurt to look directly at it. As I got closer, I had to look away because it began to make my eyes tear up. I hoped that it wouldn't be difficult to look at things on the planet's surface. A few moments later, I had my answer. The surface was bright due to snow that covered the ground as far as the eye could see. Nonetheless, it wasn't hard to look at; in fact, it was rather tranquil. The atmosphere of the planet created a darker blue appearance than that of Earth's blue sky. I imagined that Uganthos had something to do with it. It looked as if it was twilight, or that a large storm was looming overhead. The glowing archway lit up the snowy ground for quite some distance beyond it.

Looking all around the area, there were only two things to see, other than the archway that is. One was a pole that stuck out of the snow. It was about the size of a walking stick. Not knowing what its true purpose was, I used it as a walking stick. The other object was a huge cone-shaped building about a tenth of a mile away from me. I am glad that I have the energy field protecting me this time. The idea of walking a tenth of a mile in fridged temperatures did not sound appealing. The place sort of reminded me of a desert of snow, instead of sand. Instead of dying of dehydration and heat exhaustion, it would be eventual hypothermia.

Trying to be more optimistic, I head off for the structure in search of my key. The snow proved harder to walk in than I thought it would. In some areas, I sunk into the snow up to my knees. The pole kept me balanced pretty well; I'm glad it was placed there for the taking.

As I attempted to walk, I remembered the area of shimmering sky on the planet Dremida. I stopped and looked back to see if such a shimmer could be seen here as well. Although not as easy because the sky was darker, I could definitely see the distortion in the atmosphere. It was so cool to see.

Having returned my gaze to the ground, I saw that the structure was much closer. Continuing on, every step I took was an arduous task. It was hard to do better than twenty feet in a minute. Things went annoyingly slow and uneventful for the first half hour. Just then, the pole almost fell out of my hands as it went straight through the snow. I barely grabbed hold of it before losing it all together. Daring not to move an inch further, I poked the pole all around me. An area off to my left had solid ground for about five to ten more feet, and then, nothing. I stopped to think of how fortunate I was to keep the pole out in front of me as I walked. Had I kept it at my side, I'd probably be dead right now. I guess the pole's purpose was meant to be a walking stick, after all.

I couldn't go any further without knowing where to step. I, therefore, looked around to see if there was any way to get the information I needed. It took about five to ten minutes, but I was able to discover the location of a buried rock of decent size. It wasn't easy to move it due to being frozen to the ground, but I eventually got the pole under it enough that it gave way. Carefully walking back with a rock in one arm and a pole in the other, I prepared to throw the rock as far as I could. I hoped that by throwing the rock, I would receive the information I needed.

Dropping the pole, I gripped the rock in both hands and gave it my best throw. After my Olympian effort, I stood there in wonder. The rock fell straight through the snow, leaving a hole bigger than the rock should have had the ability to make. Areas around the hole showed many cracks that grew steadily larger until the entire area in front of me collapsed away.

Staring in startled amazement, I saw a beautiful crevasse before me that dropped hundreds of feet. The walls of the crevasse were made of years of compacted ice that looked a shade of blue. I was standing only a few feet from the edge and quickly moved back. I knew the walk to the structure looked too easy to be true.

I had faced a few near-death events during my adventure, but for some reason, this one bothered me in a way that I didn't understand. It really didn't make sense, but I couldn't deny my feelings of uneasiness. Trying to refocus, I picked up the pole and continued to make my way to the cone structure. The crevasse took me between two and three hundred feet off my original course, but it was necessary in order to get around it. I will go on to say that I met up with one more problem in the form of a deep-frozen pool. It, too, was hidden beneath the snow. The pole made contact with the icy surface before breaking through the ice a few moments later. Had it taken any longer to break through the ice, I fear that I may have walked out onto the ice and fallen through.

This place, despite its beauty, was deadlier than a pit of asps. Thankfully, avoiding another icy demise, I continued the remainder of the way to the building without further incident. I stood and thought of how much I really didn't like this place, and still, I couldn't place my finger on the reason why. Maybe it was the snow. I've never been a big fan of snow. Maybe it was almost dying twice, but I've already dealt with that numerous times in other locations. No, what I thought it really might have been was the feeling of isolation and loneliness. Up until going to Varyn, I've had someone or something to talk to. I missed hearing SAROS tell me about my destinations, and my new friends welcoming me to their world.

I felt depressed. I was lost in my thoughts as if trapped between two worlds. I found that I could not focus well on any one thought. Nonetheless, I turned and entered into the structure. The inside was he same as the other cone structures, with my destination about forty feet over my head. Looking at the turbo lift, I didn't even bother going over to it in the hopes of putting it to use. The transport was leaning outside the shaft, rather heavily damaged. It seemed to have experienced the same fate as the lift on Targil.

It looked as though the staircase would win this time. Thinking about climbing a long, snowy set of stairs didn't improve the way I was already feeling. Defeated, I slowly approached the stairs when I remembered the grappling hook gun in my pack.

Yes! I won, after all! I thought. *I didn't have to climb the stairs.*

Happily, I took out the gun and carefully aimed it at the walkway. The hook latched itself to the walkway about five feet to the right of the table holding the metal box. Pressing the retracting mechanism on the gun, I took a ride up to the walkway. Grabbing hold of it, I hoisted myself up onto my feet.

There it was! It seemed strange looking at this box. I felt as though I had gone through so much to get to it. I could tell that I was beginning to feel sorry for myself, so I opened the box and took out its contents. This time, it was the spiral key; its design added to its incredible beauty. Storing the key away, I head back via the staircase. Like I said once before, at least gravity would be helping me down. Walking those blasted stairs took about four minutes. When I finally got off of them, I thought of my trip back to the archway. At least I would be able to follow my own tracks back. Still, I would make sure to bring my trusty pole.

When I got outside, I saw that it was snowing. *Fantastic!* If I was going to follow my tracks, I had to hurry; they were already half buried. The snow came down pretty hard, but I managed to get past the areas of greatest danger. I soon found myself in sight of the glowing archway. Never did a sight look as welcoming to my eyes as that arch did at this time. I walked the remaining distance to the archway without any problems; in fact, it was rather peaceful. Still, I had no desire to stay on Yixelex once I reached the arch, so I walked through without hesitation. It was wonderful to see the beauty of the stars once again. Feeling relieved to be away from Yixelex, I closed my eyes while I traveled back to Varyn. Soon enough, I found myself standing in the maze room once again.

Closing the door to Yixelex, I walked over to the twelve o'clock door, pulled out the spiral key from my pack, and placed it into its rightful place on the door. It would have been really nice to hear someone talk to me for a couple of minutes, but that wasn't going to happen. Well, I suppose that isn't true. I have been hearing from Varyn. It was better than nothing.

Pulling myself back together, I took a deep breath and made my way over to the door in the four o'clock position.

Reaching over, I touched the lit rectangle. A sight unlike any other so far met my eyes. The planet before me was beautiful: a large orb of bluish-green with white swirls flowing through its atmosphere. Taking a step closer to the threshold, the voice of Varyn spoke the following:

"Hello, traveler! My name is Varyn Keb. I welcome you to one of the portals of the worlds. The beautiful planet you see before you is called Artysha, the sixth planet from Uganthos. This planet measures 22,958.7 Safilons in diameter, with an average surface temperature of -208 Laemids, making the surface practically impenetrable to instrumentation. While on Artysha, you will be protected by an energy field that will surround your entire body. As you move, it will move with you. Now, you only need to step through the portal, and you'll be on your way. Enjoy your visit!"

Hearing another voice, even a recorded one, brought a certain degree of comfort to me. *All right,* I thought. *This is enough! I just needed to focus on my task at hand.* I had two more keys to find and a mystery door to open. Feeling a bit more determined, I took my step closer to Artysha.

CHAPTER 43
ARTYSHA

Preparing myself for another challenge, I closed my eyes to meditate. After stepping through the portal, I breathed deeply feeling no sensation of movement. When I opened my eyes, I found myself standing at the base of an ice cliff with the archway behind me. Looking around from my position, I could just see the pinnacle of the cone building peeking above an overhang of snow and ice. Although I could tell that it was daytime, the amount of light was no greater than that of deep twilight. Uganthos was not a major source of light and heat from this distance. In fact, I was able to see a good number of stars in the sky.

Other than the archway, I could see no other objects of interest, not even a pole to navigate through the snow. I took that to be a good sign. It looked like I just needed to find a way over to the structure. The face of the cliff was steep and icy. At the present temperature, no item in my possession would be able to dig into the ice. The only area on the cliff that showed any promise was a small ledge some 15 feet above me. The edge of the ledge had exposed rock, enough that the grappling hook would be able to latch onto it if shot accurately enough.

As I reached to take the backpack off of me, my eyes caught sight of different colors. At first, I thought I was seeing things, but the colors were real. Getting a better position, I saw crystal formations on the ground and embedded into the cliff face. The colors were emanating from them, cycling through a set sequence. Taking out my notebook, I wrote down the order of the colors as I saw them. I had to watch the order of the colors a few times because I noticed a brief pause periodically. It must have been a starting point. In fact, there were two pauses. One was about three seconds long, the other about five seconds long. This threw me off for a while, but I soon understood as I continued looking for the structure.

After feeling confident that I had the order of the colors written correctly, I finished what I started and grabbed the grappling hook gun. Looking up toward the ledge from my new position, I saw an opening in the cliff face. From atop the ledge, the opening could be reached. Walking directly beneath the ledge, I took careful aim for the rocky exposure. The first couple of times did not yield good results; the hook refused to latch onto the rock. My third attempt was successful. As the old saying goes, the third time's a charm. Making sure the hook was secure, I held on tight and hit the retract mechanism on the gun. I was carried up to the icy ledge, only to find myself dangling there for a while. The hook was not the only thing that had difficulty grabbing hold of the rocky area of the ledge. It took about a minute, which seemed more like ten, but I eventually was able to pull myself onto the cliff's edge.

Once on the ledge, I collapsed with exhaustion for a few minutes. I laid there on my back, looking up into the sky and also at a magnificent display of incredibly majestic icicles. I couldn't believe the size of them. They had to range in size from fifteen to twenty feet in length. The largest of them had a base with a circumference of at least three to four feet. The view was beautiful but startling. They hung right over the entrance to the opening in the cliff face. I wasn't even certain if the opening led anywhere. Rising to my feet, I took a deep breath while stretching a bit. I prepared myself to move onward. Lining up with the opening while keeping away from the icicles, I could see more colors within. Wanting to make sure the icicles weren't a death trap, I picked up a loose piece of ice from the ledge and gently lobbed it into the opening. If something was going to happen, I didn't want it to be me that it happened to. The icicles remained motionless. I could almost feel them looking at me and saying, "Dude, what's your problem?" Nonetheless, I still entered the opening very cautiously, trying to pass in between the icicle's points as much as possible.

Once inside, I could see that I was in a man-made tunnel. What also gave it away was the colorful staircase that was in front of me. It was a set of fourteen stairs, each one an array of colors. On the left side, the color was red. The next color to follow was orange until it completed the visible spectrum of light. The only addition was a white-colored segment after the violet on the far right. Each of the fourteen stairs was identical in appearance. I had no choice but to ascend the stairs; it was the only way forward. The stairs could not be bypassed because they extended from one side of the wall to the other. I either went up, or I went back.

I knew that a simple walk up the stairs wouldn't be possible. I tested the stairs to see what I would encounter. Finding more ice pieces lying about, I tossed one at the first stair to see what would happen. Upon contact, the area of the stair where the ice chunk hit collapsed through the stair. Obviously, this was not a bad idea. Tossing another, the ice chunk burst into flames. *WHOA! This is too much for me to handle,* I thought. Trying once more before turning back, the piece of ice just sat where it landed. Fearing it was a time-delayed trap, I watched for over a minute. The ice chunk remained unaffected.

Gathering more loose pieces, I picked them all up and threw them at the whole staircase. The result of my action had me back away while shielding my eyes. Some of the pieces burst into flames again; a few vaporized, blades shot up out of a few stairs, and others boiled away to steam in seconds. Only a few remained unaffected.

"What is so special about those pieces?" I asked myself.

The unaffected pieces lay on stairs one, five, and eight. A few moments later, my eyes widened as I quickly took out my notebook.

Of course! That was the answer.

I should have picked up on it much sooner. The ice on the first step was on the green segment of the stair. The ice pieces on stairs five and eight lay on a yellow segment and a blue segment, respectively. Looking over the color order in which the crystals flashed, the first color was green, the fifth was yellow, and the eighth was blue. The order of the colored flashes was the answer to safely climbing the stairs. Ingenious!

I had to mentally prepare myself for making the ascent; I saw what one bad step could do. With my notebook in hand, I carefully stepped on the green segment of the first stair. The remaining colored stairs were red and then white, which was a very difficult leap, mind you. After the white, I went to blue, then yellow, violet, a tough jump to orange, then blue, then back to orange, then green, violet, a difficult jump to red, then yellow, and finally white.

After completing my jumping exercise, I stepped off the stairs feeling quite warm. It was a bit tense knowing that one bad foot placement could cost me my life. I was very relieved when I stepped off the stairs.

Continuing onward, the path led into a tunnel similar to the one I first entered into. The pathway led outside onto another cliff ledge. This time, the cliff's edge was connected to the entrance of the cone-shaped structure. There were more monster-sized icicles hanging over the entrance. I don't know why they freaked me out so badly, but they did. Thankfully, like the last set of icicles, these did not hinder my progress either.

Entering the structure, everything was designed the same as the other cones. Looking around, I saw that the turbo lift wasn't going anywhere anytime soon. On top of it was an icicle that had broken loose from the pinnacle of the cone. This momma bear made the ones hanging over the entrance seem like little toys in comparison. Its base had to be six or seven feet in diameter. I guess they can fall after all. It came straight down and completely mangled the lift. There was little left of it that was recognizable. On the icicles way down, it also had no choice but to wipe out a good portion of the walkway. Unfortunately, the table and the metal box sat on the walkway in the middle of the room. There was simply no walkway left except for the far edges.

What I was looking for was now on the ground. I looked all over the place but found only pieces of the walkway here and there. After a little more searching, I found table fragments under some of the icicle, which had broken free of the ceiling. The impressive remains now lay scattered about the floor. There was plenty of broken ice lying about, looking like uncut diamond and quartz pieces. Extending my search to look under the ice debris, I found the metal box wedged between the main icicle and some of its shattered fragments. The box had taken some damage, but fortunately wasn't anything that prevented me from opening it.

This time, it was the pyramid key that lay inside. It looked just as astoundingly beautiful as the other keys I had found. Placing the artifact in my pack, I made my way to the door. I then realized that I got out of having to climb the stairs once again. *Thanks, Mr. Icicle!*

With a smile on my face, I exited the building onto the cliff's ledge. I then thought of the colorful staircase, which now had to be done in reverse order. This thought wiped the smile right off my face again.

Sparing you the details of my return walk down Rainbow Road, it sufficeth me to say that I did not vaporize, get impaled, or burn up.

When I exited the cave, I returned to my grappling hook gun, which I foolishly forgot to take with me. Before leaving, I made sure it was still secured to the rocky surface. I then scrambled off the ledge, hanging on to the edge of the cliff with one hand and the other on the gun. I began my descent to Artysha's surface. About ten feet from the ground, the hook slipped off the ledge, causing me to land flat on my back. It was a hard landing that knocked the wind out of me, but aside from that, I walked away unscathed. Remembering my gun this time, I stored it away and headed back to the archway. Artysha wasn't such a bad place; I liked it a lot better than Yixelex.

Stepping through the portal, I took in the beauty of the stars once again. I wished to be able to continue doing this for the rest of my life. It was a nice wish, but I soon found myself standing in the maze room, looking at the twelve o'clock door. I walked over to it and placed my newly found key into it. Seven keys installed; one to go. The six o'clock door was the only one left to explore. I thought of the fact that the six o'clock door was where I entered the room in the first place.

"Won't it lead me back to where I was, I thought, *or did it change to a new destination?"*

Touching the light, the door opened to reveal the beauty of space with a violet-colored planet in the distance. I was floored that there was a new location beyond the door.

"How is this possible, I said, and what's with a violet planet?"

It was amazingly beautiful, to say the least. Triggering Varyn, he spoke the following words:

"Hello, Adventurer! My name is Varyn Keb. I welcome you to one of the portals of the world's. The incredible planet you see before you is called Orbsolox, the seventh and final planet from Uganthos. This planet is the system's smallest, measuring 3,398.2 Safilons in diameter. The average surface temperature of Orbsolox is -437 Laemids, making the surface totally impenetrable to instrumentation. While on Orbsolox, be prepared for the possibility of abrupt changes in the weather. Many reports have been given of freakish conditions that come about all of a sudden. It will also be wise to carry a source of external illumination for your excursion due to the planet's extreme distance from Uganthos. Be comforted to know that you will be protected by an energy field that will surround your entire body while visiting Orbsolox. As you move, it will move with you. Now, you only need to step through the portal, and you'll be on your way. Enjoy your visit!"

After that speech, I didn't know whether to be excited about my visit, or to be scared to death. I checked to see if I remembered to bring a flashlight with me when I left the Lotania. Thankfully, I had it with me. Taking my step into infinity, I peacefully got pulled toward the distant violet planet.

Violet! That was so wild.

CHAPTER 44
ORBSOLOX

My moments of traveling through space never lasted long enough. Words simply cannot do justice to the feelings I experienced. Approaching Orbsolox was like no other planet. Its violet color was captivating to look at. As I got closer to the surface, I seemed to have passed through a layer of atmosphere of some type. I had no idea how a planet that far away from its Sun, and being as small as it was, could have an atmosphere in the first place. It just wasn't natural. Nevertheless, after passing the barrier, the violet appearance was gone, leaving the normal sight of snow and ice.

Once on the surface, I saw that the sky had absolutely no hint of daylight to it. Uganthos had the apparent size of a ping-pong ball. SAROS never got around to telling me the distance of the seventh planet from the sun, but from the look of things, I felt like I was on the fringes of occupied space. The archway gave off a little bit of light, but it would not serve me for long. The cone structure was about two hundred or so yards away across a chasm. The cone was in the shadows, out of reach of the archway's light. I'm glad about Varyn's information concerning the need for external illumination.

As I said before, the idea of weather patterns occurring on Orbsolox was still mystifying. With a temperature of -437 Lamoos, or whatever they are, atmospheric conditions would normally not be possible. Any atmosphere should have been frozen to the ground. This place was anything but normal. I don't mean to give the idea that there was anything supernatural about the place; it just seemed to defy the laws of nature as I understood them, which was obviously very little.

Determined to retrieve the final key, I made off for the structure. When I reached the chasm, I almost turned around. The chasm was so deep, I couldn't see the bottom of it. It was just blackness beneath me, as though there was no bottom to it at all. It was like staring out into space again, only this time, without any stars.

The pathway across the chasm was nothing more than an icy bridge about the width of two planks of wood. There were no handrails to hold on to. It was just an icy plank stretching over a seemingly bottomless expanse.

As if things weren't challenging enough already, the winds got stronger as I approached the chasm. Trying not to think that I may not make it beyond this point, I considered the best way to cross the bridge. Every now and then, the wind would gust, blowing loose particles of snow and ice all around. This lessened my visibility a bit, and conditions did not improve until the gusting stopped.

The only way across the chasm was to walk the plank. Such words meant doom to the person doing the walking. Regardless, I had to try. It wouldn't have been too bad if the plank hadn't been so long. It had to be at least thirty yards in length. I sure could have done with those weighted shoes with cleats from the Lotania. Unfortunately, they were more than half a solar system away at the moment.

Focusing straight ahead of me, I began my tightrope act. I didn't get far when the wind began to gust again. It seemed to be gusting quite regularly, as if timed. During the gusts, I remained stock still to maintain my balance. I was having a challenging enough time without the wind. When the winds subsided, I slowly began to inch forward again. Everything was going great until I got a little ways away from the other side. With about five to ten yards to go, the winds became stronger than they had been. It was as if the planet was trying to prevent me from making it across. At the distance I mentioned, a sudden and powerful gust of wind made me lose my balance. I fell to my knees as I slid off the bridge. I was very fortunate to have wrapped my arms around the bridge as I slid off. Knowing that my chances of getting on top of the bridge again were practically zero, I finished the distance by walking across with my hands. I felt like I was making my way across a set of monkey bars at a playground. The only difference, though, was if I fell off the monkey bars at the playground, I wouldn't die.

When I finally reached the cliff wall on the other side, I found some decent footing and was able to hoist myself upward. Again, I found myself on my back, totally exhausted with my heart pounding, staring up at the stars. They were so bright, and the air so clear, it seemed as though I could reach out and touch them. I laid there for a few minutes enjoying the peace and rest, but unfortunately, it was time to get up. I looked back a moment at the chasm that almost claimed me, then slowly turned toward the cone, dreading the time I'd need to give the abyss a second chance.

Completing the distance to the structure, I entered to find everything the same inside. I was amazed to see that the turbo lift was still operational here.

"How can that machine operate under the present conditions?" I wondered.

Needless to say, I was quite uneasy about jumping into the lift. I just had a bad feeling, especially after seeing the lift in the last structure. I thought it best to test the lift first. If it performed adequately, I ....*might* possibly consider using it. Making my way over to it, I touched the panel to make it go up without getting inside. It took off as expected but continued to accelerate. The controls began giving off sparks. Faster and faster, it moved as it ascended. It kept on accelerating until it literally flew out of its chute, crashing into the underside of the cone's pinnacle. The collision crushed half of the lift on impact like an accordion being squeezed. The lift ricocheted off the ceiling in pieces. I had to quickly find a place to shield myself from its debris.

It took a while for things to settle down because the lift was not the only thing that fell. Like the effects of a cave-in, two huge icicles also broke free and crashed to the ground. One of them crushed part of the stairs after it fell, and the other just rammed straight into the ground like a stake, not far from the stairs. When it was safe to come out, I thought of how happy I was that I didn't take the lift. On this adventure, there is a rule that I've made. If you're on an alien world with a turbo lift, AVOID IT! This planet was dangerous! It reminded me of that rotating room in the rhombus building on Marathrax. That place had it in for me, too.

Well, it had to happen. The stairs finally win one. It seems that I have no choice but to climb the staircase. Grumbling a bit under my breath, I began to ascend the stairs. The blasted things wrapped around the wall twice before reaching the walkway, and what made it worse was the area that was blown away by the falling icicle. This was the definition of adding insult to injury.

The stairs remained intact for about a third of the way before being crushed to rubble. The area of the stairs that continued upward was too far out of reach to jump to. The grappling hook gun was of no use because of the incredibly low temperature. The ice was impenetrable under the conditions. The only thing available to me was very risky. The icicle that crushed the stairs could be used as a stepping stone, so to speak. When it hit the stairs, a piece of its top broke off, leaving an area large enough to jump on to. That part of the icicle was higher than the stair I was on. It would be reachable if I jumped hard enough. The broken area of the ice formed a crude staircase of its own. From its highest point, another decent jump would put me back on the stairs again.

Determined to give it a try, my jump from the stairs to the icicle was a little weak, and I had a hard time gripping the icicle to pull myself up. I fell about ten feet, but thankfully, the acceleration due to gravity on Orbsolox allowed for a slower descent and a milder impact on the ground. After brushing myself off, I went back for a second try. This time I was successful in all of my jumps. I found that the remainder of the staircase was okay, except for one more gap that was fairly easy to jump over to.

Once at the top of the stairs, I walked over to the metal box and retrieved the cubicle key. This beautiful item completed the set of keys needed to open the 12 o'clock door. I placed the key safely in my pack and was now ready to head back. I was turning around toward the stairs to retrace my steps when I saw a fairly sizable hole in the wall a little ways past the table. Going over to it, I looked into the hole and saw that it led outside. At first, I was startled by the sight of the drop, but on a detailed second look, I noticed a steep but manageable slope of icy snow that went to the ground. I decided to take a ride rather than go back down the stairs.

The slide was a bit harsher than I figured it would be, but I made it down uninjured. By the time I reached the bottom, I was moving at a pretty good clip. At the bottom of the incline, I continued to slide for some distance. I also did quite a bit of uncontrollable rolling, which left me with a number of bruises. It seems I would do anything to avoid stairs. Ouch!

Slowly and painfully, I rose to my feet. Hobbling along, I began my way back to the arch. My walk back to the icy plank took a while due to my painful stiffness, but I eventually arrived. Staring at that pathetic excuse of a bridge, I tried to figure out the best way to make it across. I thought of ways to lessen my chances of losing my balance. *That's it! Balance!* I needed to change my center of mass. I quickly got down on all fours and laid face down on the bridge. Wrapping my legs around the path, I gripped the sides with my hands. In this position, the wind had very little effect on me. It took me quite a while to shimmy myself across the chasm, but I made it without incident. Rising to my feet again, I saw the glowing archway in the distance. Once again, I found the archway to be a very welcoming sight. When I arrived, I turned and took one more look at Orbsolox before stepping through the portal.

Once through, I watched as Varyn got steadily closer, hoping that someday I'd be able to glide amongst the stars again. Before I knew it, I was standing in the maze room. I took one more look at the beautiful violet orb before turning away. It was time to see what lies beyond the 12 o'clock door. I figured it must be something pretty special to require eight keys to open it.

Once at the multi-locked door, I inserted the cubicle key. The rectangular panel illuminated, which made me start breathing a little quicker for some reason. It was as if I got nervous all of a sudden. Lowering my arm, I touched the panel, and the door slid open. In front of me was a tunnel that measured about 10 to 15 feet in length. Beyond it was a well-lit room that looked very spacious. Stepping inside, I looked forward to investigating this well-protected domain.

CHAPTER 45
THE LAIR OF THE ZYNDOK

Once inside, the door closed behind me, leaving me with that unwanted feeling of entrapment. The tunnel seemed to have been hewn out of the rock. It must have taken some time to dig out ten to fifteen feet. As I walked toward the room ahead, I saw a message written on the right-hand wall of the tunnel. After translation, the message read:

"Remember the test and why you are here; be kind and generous, and do not fear."

These words were familiar to me. I then remembered it was because I had read them before. Here on Varyn, in the room with the color puzzle, the words were written on the right-hand wall of the room. I still had no idea what the words meant. They spoke of a test, my reason for being here, and of being generous. I stood and pondered for a minute before continuing. A little farther down the path, I came across additional writing. When I translated it, I read:

"Walk brave and true, be full of care, for you have entered the Zyndok's lair."

"A what?" I said. "A Zyndok!"

Do I even want to know what a Zyndok is? I thought."

Looking on both sides of the remaining tunnel walls, I saw no further writing. When I reached the end of the tunnel, my mouth hung open a slight bit. I remember saying the room seemed spacious while at the entrance. Well, the room wasn't just spacious; it was enormously spacious. Looking up, the height seemed to exceed eighty feet easily. It must have spanned all the distance I traveled downward to get to this point. The room had to be on the order of fifty feet in length and width. There were different levels to the place. Two pathways could be seen beyond a pillar that must have been at least fifty feet tall. Scattered throughout the upper region of the cave were stone-looking objects against the walls. It could have been my imagination, but I can swear that I saw one of them move. I had a bad feeling about whatever those things were.

I couldn't get over how high this room went up. I didn't realize that I had actually descended so far into Varyn by having passed from room to room. The whole cave looked pretty much the same everywhere. All of the walls were made of rough reddish-brown stone. To my right, about thirty feet away, was the huge pillar. I wasn't sure, but I thought I saw a glow of some sort coming from the top of it. Climbing the pillar didn't look safe, or even possible, for that matter. Despite the rough appearance of the rock, there didn't seem to be any decent hand and foot holes for climbing. I was about to pull out my grappling hook gun when I noticed, off to my left, an object lying on a ledge about four feet off the ground.

Walking over to it, I saw what looked to be a jetpack, you know, the kind you see in older movies. It had two controls to hold on to while wearing it. It also had a red button for starting and stopping. Picking up the device, I slipped it over my shoulders. The pack also showed that it was fully fueled. Trying my luck, I pressed the red button, and the pack began to lift me off the ground. Experimenting a little, I learned that the control stick on the right chose the direction to be traveled. The control on the left side handled the velocity. Pushing forward sped me up, and pulling back slowed me down. The pack took a little getting used to, but I eventually got the hang of it.

Soon, I found myself standing on top of the fifty-foot pillar. I'm glad that heights didn't bother me too badly. As I approached the pillar's center, I saw the two extending paths more closely. One of the paths continued off the pillar from the 2 o'clock position, as seen from my position. It led upward and curved around to the left in a U-turn sort of way. The path went up an additional 30 feet or so and then leveled off on a flat, spacious rock. The other path extended from the pillar from a 10 o'clock position, as seen from my position. It stayed level with the top of the pillar. The pathway extended outward about 35 feet and came to an area of flat rock just outside the mouth of another cave. From the ground, the cave was not visible due to an outcrop of rock on the ledge. The cave entrance had to be at least 15 feet high, arching around as it came down on both sides with a width of eight to ten feet.

At the time, I had no idea that this cave was the home of the Zyndok mentioned on the tunnel wall. I may be getting a bit ahead of myself, but I feel the need to give a description of the Zyndok. It was a sight that I will not soon forget. It was a creature of immense stature, standing nearly twelve feet tall. It appeared to be made of the same rock that the cave was made of. Its body was covered with rhinoceros-like horns, some over a half a foot long. It looked so much like the cave walls, if it stood still, it would blend in perfectly. Despite the fact that the creature was enormous and made a cactus' thorns look like a man's five o'clock shadow, it was the eyes of the creature that chilled me to my very center. I will never forget those eyes; they were solid white, and felt as though they could rip the very soul from you.

OK! It's time to think of something else.

As I mentioned before about the pillar I now stood upon, I thought I was able to see a glow coming from the top. It turns out that I wasn't seeing things. About five feet in front of me, there were two small columns about two feet apart, each standing four feet high. The pedestal of each column was about one square foot. On top of the left column was a clear, smooth sphere of flawless diamond measuring eight inches in diameter. This item was no doubt priceless. It was from this diamond sphere that the glow was being generated. Within the sphere, I saw many arcs of energy moving around. The arcs caused the diamond orb to give off an inviting but powerful blue glow.

On the right column was a very valuable container measuring 8 x 6 x 5 inches. It looked as though it was too valuable for even a king to own. Inside the container, I found a collection of small triangular-shaped diamond chips. I had no idea what they were or what their purpose was. Along with the chips was a rather sophisticated-looking handheld device that displayed information about the Dhirniy people by way of a hologram. The device stored a vast amount of information on Dhirniy religion, mathematics, science, medicine, engineering, and so much more. All of these items were incredibly valuable and interesting. I could not wait to learn more about them from Nirton and Caynos.

To keep the three objects safe, I placed them in my backpack. A few moments after doing so, I heard the sound of something very heavy, thumping slowly. Needless to say, I turned to face the sound, which came from the direction of the Zyndok's cave. With my eyes wide and my mouth hanging open, I stood looking at the Zyndok standing in the doorway of its lair. The sight of this creature was very alarming, causing me to consider an escape. This was my first dose of seeing the Zyndok's white eyes. After staring into them for a few moments, the Zyndok began to move slowly toward me, lessening the 35-foot distance between us with every step it took. Completely terrified, I began to back up and bumped into one of the columns, nearly tipping it over. Due to the size of the Zyndok, it covered the distance between us in only a few steps. When it reached me, it stood no more than five feet away from me. I thought I was about to exit this life. Thankfully, instead of squashing me, it only stretched forth its spiky arm and opened its thorny hand without saying a thing.

I stood there looking at it like a deer, staring into oncoming headlights. I practically wet my pants, wondering what it was doing. It stood there waiting as if wanting something. *Of course!* The items I just took from the columns! The Zyndok must be the protector of the objects.

Panicking, while going over my options, I decided not to press my luck by sticking around. I, therefore, ignited the jetpack and flew back down to the ground in the hopes of leaving this room. By the time I reached the ground, the Zyndok was already more than halfway down the pillar. Getting ready to fly back up while the creature was climbing down, I noticed that the jetpack's fuel gauge now read half empty. It couldn't be helped; I had to stay ahead of that monster. I ignited the pack again and flew back to the pillar. The Zyndok then began to ascend the pillar again. I saw that the jetpack was down to a quarter of a tank now, with the low fuel light blinking at me. I couldn't keep this up. If I flew down one more time, I'd be out of fuel with a behemoth on my tail. I'd be trapped on the ground. I'd have to use the tunnel to hide in and, if necessary, escape the room the way I came in.

I, therefore, took my last trip to the ground as the Zyndok neared the top of the pillar. As expected, the beast immediately descended the pillar again. I couldn't tell if the Zyndok was angry or not. Its facial features never really changed. At the very least, I could only imagine that it must have been annoyed with me. Not wanting to wait around to find out, I ran for the tunnel as the Zyndok reached the ground. I made it into the tunnel only a few moments before the Zyndok reached it as well. I ran to the door at the end of the tunnel. I then turned to see the spiky arm of the Zyndok only a few feet away from me. The hand of the creature opened again. That thing had really long arms; seeing the tunnel's length was about fifteen feet. I was now trapped in the tunnel. There was no way past the Zyndok's arm. It filled the entire passageway. What the arm didn't block, the protruding spikes did.

My only escape was through the door I initially came in. I just had to hope I'd find a way to escape once I was back inside the maze room. Thankfully, there was a rectangular panel on this side of the door as well. Even better, it was lit up. I touched it, and the door slid open. I immediately jerked backward, slamming myself against the wall. I had to grip the wall where I could. In front of me was another view of outer space, only this time with an active black hole trying to pull me into it. As carefully as I could, I reached over to the lit panel and touched it again. The door closed with me panting behind it in total amazement.

"What was that?" I said aloud. "*What happened to all the rooms I was in?"*

Man, that got my heart rate up. I thought the Zyndok was bad, but at least it hasn't tried to eat me like the black hole just tried to do.

Turning my attention over to the Zyndok again, I noticed that it was becoming understandably impatient. It balled up its fist and slammed the ground quite hard. The tunnel began to lose small pieces from the ceiling and walls. I had to give it what it wanted right away, or I would not have a later.

I took off my pack and handed the ornate box that held the diamond chips into the Zyndok's spiky hand. I also added the handheld device as an extra gift. The creature then retracted its arm, took the box and handheld device out of its hand, and placed them on the ground. A moment later, it stuck its arm back in and held out its hand again. Looking nervously away to find the remaining object, I pulled it out and handed it over to the Zyndok. Once again, the beast's arm retracted, and the brute left the tunnel. I waited a small while to see if it would return. When it didn't, I cautiously walked to the tunnel's exit and looked around the edge of the wall. The Zyndok was still out there near the pillar. Looking down, I saw the box of diamond chips and the hand-held device. At first, I was very hesitant to touch them, but there was no way I was going to leave them sitting there. I picked up the items and quickly ran back into the tunnel again. The Zyndok did not come after me. Apparently, for some reason, it wasn't interested in the box and handheld device, only the diamond sphere.

Slowly making my way to the tunnel's edge again, I looked around the wall one more time. The giant was still there, its white eyes still looking in my direction. Just then, to my complete astonishment, I heard the creature say, "GET ON!"

It had a very low and gruff voice, which sounded like boulders being rubbed together. It was as if it knew I had no way up the pillar again. About ten to fifteen seconds after its initial offer, it said:

"GET ON! THIS IS YOUR LAST CHANCE."

Having no desire to further annoy the beast, I approached the Zyndok. It gently placed me on its shoulder, avoiding a six-inch spike. Up we went to the pillar's top, where it sat me down before leaving me in peace. I watched in awe as the creature returned to its cave. I stood there, feeling my body shaking due to a mixture of fear and excitement. It took me a few minutes to calm down enough to think clearly. When I felt better, I looked up and saw that I had a ways to go before reaching the top.

Chapter 46
Two Worlds Unite

I got to my feet and made my way to the upper level of the cave. As I walked, I was able to get a better look at those hanging rock-like objects. I wasn't going crazy when I said I thought I saw one of them move. Now, I was clearly able to see a number of them move. I didn't dare make any unnecessary noise in the fear of disturbing them and possibly being attacked. Not knowing what the creatures were, I simply ended up calling them rock bats. I stayed away from them as much as I possibly could.

Arriving at the upper level of the cave, I had a choice of either going right or left. On both sides of me, there was a door to another room. I figured that the door to my left led into a room right over the home of the Zyndok, so I chose to go to the right first. Let's just say that I needed a break from Mr. Zyndok for a while. I'm telling you, those eyes were too much. You'd know what I mean if you saw them for yourself. Let's just hope that you don't ever have to.

To my right, I saw the door about ten feet away. The door was centered between two side panels. On the left panel near the top, I saw a computer screen. Underneath the screen was a clear circular lens about six inches in diameter and a small unlit circle. On the right panel, I saw an indentation in the shape of a hemisphere or bowl. Under the indentation was a keyboard offering Dhirniy letters. From what I have learned of the Dhirniy alphabet, I could tell that the letters were not arranged the same as a keyboard on Earth.

Under the keyboard, I saw a small skinny slit; and an unlit rectangular light. Like all the other doors entered into thus far, I knew I wouldn't be going through this door until the panel was lit. Judging from the look of the bowl-shaped indentation, I'd say the object I needed just walked off in the hands of yours truly. *Fantastic! What was I supposed to do, bake the Zyndok some cookies and hope he'd trade me?* It wouldn't work anyway; I stink at making cookies.

Before attempting to retrieve the sphere and ending up losing my life in the process, I checked out the room on the left side. This door had a lit rectangle waiting to be touched, so I touched it. Inside the chamber, much to my discomfort, I found a turbo elevator built into the cave wall. Next to the elevator written on the wall was a message written in the Dhirniy language that read:

"This is a passage to Varyn's surface. Please put on your helmet before entering the airlock."

Putting on my helmet, which, quite frankly, I forgot I even had, I entered the airlock door by touching another lit rectangular light. Once inside, the airlock door closed and ascended to the surface of Varyn. When the airlock door opened, I walked out, once again setting foot on Varyn's surface. Looking around, I saw another ship accompanying mine, and its occupant was standing outside its main door. I immediately approached the other spacecraft and its pilot as briskly as Varyn's gravity would allow me. I began a conversation with an old friend.

"I can't believe it!"

"So, you got past the Zyndok in one piece, I see," responded Nirton.

"Rich, I mean Nirton, is that you?"

"Well, it certainly isn't any of your other colleagues from work."

"How long have you been waiting up here?" I asked.

"Not too long. I felt a bit worried when I forgot to tell you about the Zyndok before your departure. I figured that I would come in and assist you if you didn't come out in a reasonable amount of time."

"You mean you would come in to remove my corpse," I said.

"Well," Nirton grumbled while looking at the ground. "Anyway, you did great. You proved that the people of your world have it inherently within themselves to survive and grow. I look forward to when our worlds can unite."

"I'm sure in time it will happen," I responded. In the meantime, I hope to be able to carry on our relationship and visit often."

"I would like nothing better myself," Nirton confirmed.

"Come, I still have much to show you, and we have a whole city of people I want you to meet now that you can. The people of the city wait to honor you for the brave and kind deed you performed on their behalf. I hope you will allow them the opportunity to do this."

"That is very kind of everybody, but they really don't need to."

"They feel they do, Lee, and as their new leader, allow me again to say thank you for all you have done."

Looking at the ground, I did not know what to say as Nirton continued, "Let us go, my friend. Everybody's waiting."

"Nirton, just a minute, please, if I may. I wanted to ask you something about a device in the cave that was going to scan me. What would be the need for the machine to do so?"

"You will see shortly. Remember, I told you that I had more to show you. I'm sure you will find it very interesting."

"I may not be able to find out," I said. "The device needs a glowing energy sphere to make it function. My friend the Zyndok took the sphere from me. It wasn't in the mood to give it back any time soon."

"The sphere is called a Glydos, and the Zyndok was supposed to take it from you. It was a test designed to prove your kindness, generosity, and bravery. Remember what was written on the wall of the color room and again in full on the tunnel wall?

Remember the test and why you are here; be kind and generous, and do not fear. Remember that!

I also mentioned words similar to those as a hologram when introducing you to the Lotania. The Zyndok was only doing its job. As you have noticed, it does it very seriously. Take comfort, though, in knowing that it would not have harmed you, even if you decided not to give it back.

You need not worry, my friend; you will see the device operate. I would not dream of depriving you of that experience."

Nirton gestured with his hands toward the ship and said, "Shall we?"

I followed Nirton onto his craft and was soon on my way back to Marathrax. As the ship got closer, I saw a vast multitude of people gathered together. I began to feel butterflies in my stomach. When the craft landed, the door opened, and the sound of the crowd began to grow. When I exited the ship, a loud roar of cheers erupted. I can only imagine what my face must have looked like at that time. I slowly followed Nirton into the large crowd of people. They took my hand in the attitude of giving thanks and appreciation. Despite the din that surrounded me, I somehow heard one of the townspeople call out. When I looked up to see the individual, I saw that it was Caynos, the individual I rescued with Nirton. I then heard Caynos say,

"My fellow Dhirniy, please, if I may have your attention."

The crowd quieted down, and Caynos continued, "Thank you. As you know, the past few days have been disturbing, but they have also been very special as well. I say very special, because we have made contact with a friendly, caring, and courageous people, one of whom stands with us today. I say disturbing because we have dealt with damage to our wonderful city and, mostly, the loss of people we love. All of this has come about because of an act of violence based on fear and anger from our brothers and sisters of the world, Allorayus. Even so, we must remember not to hold anger. Also, we must not have any desire for revenge in our hearts towards our neighbors. Instead, we must endeavor to unite our two worlds as we once were a long time ago. In the meantime, we will rebuild our city and strengthen our defenses so that no such attack is possible again.

As for now, let us honor our new friend who acted above and beyond any call of duty for a people he never knew existed until a short time ago. Captain Travis, please allow me to personally extend my friendship and appreciation for all you have done."

Nirton then walked up to the stand beside Caynos and said, "Captain Travis, I want to express my feelings of appreciation and friendship. I have had the pleasure of knowing you for the past seven of your Earth years. Your friendship and acceptance toward me were greatly needed and appreciated. I knew that it was right that I introduce you to this world. Forgive me for what you have been through in order to get here, but I knew you could do it. Please, won't you come up and share a few words with us." The crowd broke out in very loud applause and cheers as I took the stand. I then spoke the following,

"I am truly overwhelmed at this time. This experience is a lot for me to take in, but I'm enjoying every moment of it. Since I was a boy, I have always dreamt of traveling into space. I'd always imagined myself traveling to different stars in the universe. I figured that my imagination was the closest I'd get to actually doing such a thing. Well, it seems that I received more than I ever expected. Not only have I made it into space due to the diligence and intelligence of a wonderful friend, but I also stand before a vast group of wonderful people from a distant world whom I know I shall stay close to. My words can never truly express the feelings of thankfulness and fellowship I have at this time. Thank you for your warm acceptance of me and the hope of uniting with the people of Earth someday. I hope that day will come very soon."

The crowd applauded and cheered as I looked at the smiles upon their faces. After a few moments, the crowd settled down and Nirton spoke directly to me,

"Captain Travis, as the new leader of Marathrax, I hope the people of this city will join me in presenting the key to our city to you."

I looked out over the crowd who all cheered loudly as I stood there in amazement. Looking back toward Nirton and Caynos, I noticed an odd look on Cayno's face as he looked at Nirton. It seemed as though something was wrong, but I didn't have much time to think about it. My attention was drawn to an object being handed to me. It was a Glydos like the one the Zyndok took from me. Nirton handed it to me with a big smile on his face. He then began to laugh as he looked at the expression on my face.

Nirton continued speaking, "Now, you may come and go as you wish, my friend. Please allow me to show you how this can be done. For me to do so, we must go back to Varyn where I will instruct you."

I just stood there looking at Nirton, unable to say a word. I gazed at him with curiosity over the words he spoke to me. He then smiled and walked with me back to his ship. I waved at the people as I walked away with Nirton. The crowd started roaring again as I entered the ship. As the door was closing, I caught sight of Caynos looking toward the ground as if deep in thought and very disturbed. When the ship's door closed, the sound of the crowd became muffled but still quite audible. As the ship lifted off, I thought of how wonderful this race of people really was. I felt as welcome as I did overwhelmed.

The trip back to Varyn happened with very few words spoken. It seemed wrong to disturb the peace I was feeling, but before I knew it, the ship was descending toward Varyn. Nirton and I prepared to exit the craft by putting on our helmets and checking our suits. Moments later, I was standing on Varyn's surface again. It seemed so long ago that I was here, when in reality, it was only about two hours ago. Nirton pointed over toward the outcrop of rock I saw when I first landed. He waved me forward and walked over to it. I checked the time, thinking it may be close to 81:75:00 rotations, but my timepiece read 50:14:26. Looking at Nirton, he seemed to have understood what I was thinking. He then said to me,

"May I please see the Glydos?"

I handed the sphere over to Nirton, who then passed it over the face of the rocky outcrop. At that time, the airlock that took me back to the surface came rising out of the ground. I then responded,

"It would have been much easier if I could have done that the first time."

Nirton smiled as he stepped aside to let me enter the airlock. After following me in, the airlock door closed, and we were both lowered into the cave below. We removed our helmets as I followed Nirton over to the scanning device I could do nothing with earlier. Nirton then said,

"As I have promised, my friend, you will finally get to see how this device works. Please insert your Glydos."

As instructed, I placed the sphere into its indentation. Once inserted, the small circular light to the right of the indentation illuminated. Nirton then continued, "The machine is ready to scan you and will commence when you touch the illuminated circle."

"What is its purpose for scanning me?" I asked.

"You will now see. Please proceed," answered Nirton.

I moved closer to the lit circle and reached over to touch it. After doing so, the machine swept a wide beam of light over me a couple of times. When completed, the familiar rectangular light illuminated, indicating that I could now enter the room. Without hesitation, I touched the panel, and the door opened. Unprepared, I jumped back a bit as I gazed out into space once again. After the initial shock, I immediately moved right up to the door threshold. I couldn't believe what I was seeing.

It was the Earth!

I was home.

Completely amazed, I quickly looked over at Nirton, who said,

"It is real, my friend. That is your home planet. When you step through that door, you'll be on your way home. It will take you back to TransQuest where this all began."

"Nirton! This is incredible! How is this able to be done?"

"By scanning you, the teleport was able to match your species with your planet."

Looking down, Nirton reached for a triangular information disk under the keypad, which the teleport ejected after my scan was completed. Nirton then continued,

"Here, add this information disk to the others within the box you picked up. This is your ticket back to Earth from whatever world you visit."

Looking totally blown away, I said, "You mean to tell me that each of these triangular disks is a way to get to another world?"

"That is correct, my friend."

"There are quite a number of disks in this box," I said.

"44 to be exact; well, 45 now," Nirton responded. "And I'll let you know that you will be most welcomed in whichever world you choose to visit. Just be sure to give each place your home disk so that their world may unite with yours, as ours has united with yours."

"Nirton, I don't know what to say. This is totally amazing. This is a dream come true. The thought of being able to travel to other worlds is mind-boggling. But how will I do it? Once I travel through this door, I'll have no teleport in which to insert a disk, or the sphere. No such device exists on Earth."

"You may call a teleport to your location anytime you wish," responded Nirton. "All you will need to do is hold a triangular information disk in your left hand and, in your right hand, pass the Glydos over the information disk. This will call up a teleport anywhere, at any time. Just be discrete about it; because as you said, such technology doesn't exist on Earth.

Oh yes, only you can call up a teleport. The disks and the Glydos are connected to your molecular pattern. Therefore, the objects are just valuable trinkets in the hands of others. Please know that when I say valuable, I mean it very seriously. Such objects will be very tempting to steal if seen. But take comfort in knowing that if the items are ever stolen from you, the offender(s) will be rendered unconscious after touching them. They won't get far. Nevertheless, take excellent care of them, and please don't lose them."

"You can count on it, my friend," "They will be well looked after."

"You will find the names of the worlds on the disks," Nirton said. "You may wish to organize them in a manner which best suits you."

After a few moments of looking at the disks, Nirton continued, "Welcome to the family, Captain Travis."

"Thank you, Nirton. You can't imagine what this means to me. I'm going to miss calling you Rich and having you around the workplace. What do I tell the others when they ask where you are?"

"They hardly ever paid any attention to me when I was there, except to laugh at me, of course. Should they actually ask about my absence, just tell them that I went home to be with family. Let them know that I will not be coming back. I don't suppose many tears will fall."

"Some tears will," I said.

"Thank you, my friend. That means a great deal to me.

You now have the means to visit anytime you wish. Please come often. There are still many things I wish to talk to you about, and many places and things I want to show you. I'm sure that you will find it all very interesting."

"There's still more?" I said.

"I told you before that there are many more things I wish to share," Nirton responded.

"Yes, but I thought you meant the teleport device."

"That was part of it," Nirton said. "But there is still more for when you visit again."

"I look forward to my next visit already."

"As do I. Until then, my friend! Take care of yourself."

"And you, Nirton. Please say goodbye to Caynos and the people of the city for me."

"Oh, and by the way, I continued, You may want to check up on Caynos; he seemed rather disturbed when we left the group."

"I'll be sure to do that. Thank you for letting me know."

"Goodbye for now."

"Goodbye."

I then turned toward the amazing view of my home world. It is true that words are unable to describe the immense beauty of Earth. Someday, I hoped many from Earth would have the chance to experience a moment like this. Before stepping through the portal, I looked back to see Nirton smiling. Acknowledging one another one more time, I turned and walked home.

The journey back was exquisitely beautiful and peaceful. I loved traveling through space without a spacecraft. It was no doubt one of the best parts about my adventure. Then again, there were so many incredible moments, I'd find it hard to list them all if I were asked to do so.

Soon, I found myself standing outside TransQuest Corporation. Stepping away from the building a bit, I looked up at the stars and smiled. I felt as though I was looking at personal friends.

Lowering my eyes, I was lost in the many thoughts of my recent adventure. It was mind-blowing to believe that I was given a gateway to the universe. I would soon be traveling to worlds only known to God. I wondered which world I would visit first.

CHAPTER 47
FULL CIRCLE

There goes the Sun again, preparing to close another day. I sit here watching time slip by, leaving my job progressively later each day.

It has been about a week since my return and nothing around here seems the same. I go about my job as I always have, but with a knowledge that changes how I see and do things. I miss Richard a great deal. He is the only person I can share my thoughts with now. Nobody else would understand. They would put me away if they heard me speak about the things I have learned.

"Good night Lee! Have a great evening. See you in the morning."

Yeah! You too, Phil. Have a good night."

There goes one of them right now. Don't get me wrong, Phil's a great guy, but he doesn't have an open mind. He never got along with Rich when he was here. He just couldn't get himself to believe in the company's goal. Phil basically comes to work for the paycheck.

As expected, when Rich didn't show up for work for a couple of days, people asked why. When they were told that he resigned, most of them just nonchalantly said, "Oh, that's too bad," and walked away.

As far as I'm concerned, the only reason why TransQuest excelled at all was because of Richard. He was the only one who knew what he was doing. I can understand why now, but I dare not share it with anyone.

"Lee, you're here rather late. Are you planning on going home tonight?"

"Oh! Hi Lisa."

"Yeah, I was just finishing up a few things. I'll be leaving shortly."

"Well, I hope you finish soon. Have a good evening."

"Thanks, Lisa, you too. ....Hey, Ah, Lisa!"

"Yes, Lee."

"What would you say about having dinner with me tomorrow night?"

"Wow, Lee! That was unexpected."

"Oh, I'm so sorry, Lisa. I didn't mean to be forward, or to offend you."

"No, Lee! You didn't offend me at all. I'd love to have dinner with you. Thank you for asking me."

"Really! You'll have dinner with me?"

"I'd love to, Lee."

"Wow! I mean Excellent! What do you say we work out the details tomorrow? I'll pick the restaurant. It will be a surprise."

"I look forward to it."

"This is great! Good night, Lisa."

"Good night, Lee."

It was nice seeing the smile on her face as she got on the elevator. Lisa is a wonderful lady. She was one of the few who had a genuine feeling of loss when Richard left. I look forward to knowing Lisa better. I can sure use the companionship in my life. Perhaps, she will be someone I can share my adventure with someday. That would be wonderful. Then I won't feel so alone anymore.

Well, I suppose I should call it a day. I'm just having a hard time getting myself to leave. I keep going over the events of my adventure, especially my final moments with Nirton.

As I mentioned earlier, it has been about a week since I left Marathrax. With the gift of being able to visit different worlds, you'd think I'd be anywhere but here right now. I just haven't found the bravery to show up uninvited on someone else's doorstep.

What do I say? "Hi there, I'm from the planet Earth. Can I come in?"

Rich, I mean Nirton, was certain that where ever I went, I would be welcomed. I believe him, but still.

I have looked at the diamond chips time and time again. I guess it's time to stop looking, and start traveling. Perhaps I'll give the planet, Psiordan, or Rysomthibae a try. I don't suppose it really matters. I plan on visiting all 44 of them eventually. Maybe I'll take Lisa somewhere VERY special for dessert tomorrow.

Well, it's 6:53 p.m. I'm out of here.

I rose from my chair and walked to the elevator. When the door opened, I stepped inside. I could smell the perfume that Lisa was wearing still lingering in the air. It was no doubt one of the best moments of my day.

Wishing I could leave my nose inside the elevator, I walked out of it and then left the building. Once outside, I walked over to my vehicle and entered it. Looking at the coordinate keypad inside my car reminded me of the Lotania for a few moments. Being the Captain of that craft was wonderful; terrifying at times, but wonderful.

Breaking free of my daydream, I input the coordinate 007264, and received an automatic ride to my place of residence. Eight minutes later, I was there.

Home at last!

The End

Our Sun's Family
(The Book on the Secretary's Desk)

Name	Size (mi.)	Dist. (mi.) (Prim.)	Escape Velocity (mph)	Rotate Velocity (mph)	Mass (kg)	Temp. (F) Avg.	Surface Conds.	Major Atm. Gas	Gravity
Sun	864,956	-----------	1.382E6	4,470	1.99E30	9,941	----------	H, He	High
Mercury	3,033	3.598E7	9,509	6.87	3.30E23	332	Rocky, Cratered	None	Low
Venus	7,521	6.724E7	23,184	4.06R	4.87E24	867	Rocky, Cratered	CO_2	Normal
Earth	7,926	9.296E7	25,008	1,040	5.97E24	59	Water, Rocky	N, O	Normal
Mars	4,221	1.416E8	11,236	539	6.42E23	- 82	Rocky, Cratered	CO_2	Low
Jupiter	88,846	4.837E8	133,098	28,123	1.9E27	-163	----------	H, He	High
Saturn	74,898	8.867E8	79,411	22,081	5.68E26	-219	----------	H, He	Normal
Uranus	31,763	1.783E9	47,647	5,788R	8.68E25	-323	----------	H, He, CH_4	Normal
Neptune	30,775	2.795E9	52,568	6,001	1.02E26	-330	----------	H, He, CH_4	Normal
Pluto	1,477	3.67E9	2,707	30.27R	1.30E22	-381	Rocky, Cratered, Icy	CH_4 (Frozen)	Very Low
Luna	2,160	238,855	5,316	10.38	7.35E22	240	Rocky, Cratered	None	Low
Phobos	14.2	5,827	25	5.67	1.07E16	- 40	Rocky, Cratered	None	Extrem. Low

Deimos	7.71	14,579	12.4	0.81	1.48E15	- 40	Rocky, Cratered	None	Extrem. Low
Io	2,264	262,032	5,719	167.02	8.93E22	-262	Rocky, Volcanic	None / Vol. Emi.	Low
Gany-mede	3,274	665,116	6,132	59.87	1.48E23	-261	Rocky, Cratered, Icy	None	Low
Europa	1,940	416,878	4,530	71.91	4.80E22	-276	Icy	None	Low
Callisto	2,995	1.17E6	5,460	23.39	1.07E23	-219	Rocky, Heavy Cratering Icy	None	Low
Titan	3,200	759,235	5,908	26.27	1.35E23	-291	Rocky, Cratered, Icy CH4	N, CH4	Low
Rhea	949	327,530	1,421	27.55	2.31E21	-319	Rocky, Cratered, Icy	None	Low
Titania	980	270,862	1,698	14.74	3.40E21	-334	Rocky, Cratered, Icy	None	Low
Miranda	293	80,399	425	27.25	6.40E19	-352	Rocky, Cratered, Icy	None	Very Low
Triton	1,682	220,437	3,255	37.52	2.14E22	-391	Rocky, Cratered, Icy	None	Low

Charon	753	12,174	1,320	15.30	1.59E21	-364	Rocky, Cratered, Icy	None	Low
Iapetus	913	2.213E6	1,282	1.50	1.81E21	-262	Rocky, Cratered, Icy, Sooty	None	Low
Oberon	946	362,583	1,642	9.20	3.08E21	-325	Rocky, Cratered, Icy	None	Low
Nereid	222	3.42E6	340	60.16	3.1E19	-370	Rocky, Cratered, Icy	None	Very Low
Ceres	584	2.574E8	1,154	203.55	9.38E20	-158	Rocky, Cratered, Icy	None	Very Low
Comet Halley	6.84	1.649E9	4.5	0.41	2.2E14	-267	Rocky, Icy	H20, CO2, CO	Extrem. Low

Astronomical Formulae

Escape Velocity = $(2GM / R)^{.5}$

Where G = 6.67259E-11 Nm^2/kg^2 (Gravitational Constant of the Universe)

M = Mass of the body being escaped (kg)

R = Radius of the body being escaped (meters)

.5 = Taking the square root of the information within the parenthesis

Impact Time = $(2h / a)^{.5}$

Where h = initial height of the falling body (meters)

a = acceleration due to gravity (m/s^2)

Gravity = 4π^2L / p^2

Where L = length (meters)

p = period (seconds)

π = Pi (~ 3.1415926535)

Roche's Limit = 2.44 $(\rho M / \rho m)^{.333}$ (R)

Where ρM = The larger mass' density (kg/m^3)

ρm = The smaller mass' density (kg/m^3)

R = Radius of body (miles)

.333 = Taking the cubed root of the information within the parenthesis

Surface Temp. = $[L(1-a) / 16\pi d^2 e\sigma]^{.25}$

Where L = Star's Luminosity (Watts)

a = Body's albedo

d = Body's distance from the star (meters)

e = Body's emissivity

σ = Stefan-Boltzmann constant (5.6696E-8 w/m^2)

.25 = Taking the fourth root of the information within the parenthesis

Gravity Law (F) = GMm / r^2

Where G = 6.67259E-11 Nm^2/kg^2 (Gravitational Constant of the Universe)

M = Mass of the larger body (kg)

m = Mass of the smaller body (kg)

r = distance between the two masses (meters)

Angular Size (α) = d (206,265) / D

Where α = Angular size of body (² of arc)

d = Diameter of body (miles)

D = Distance between you and the body (miles)

Density (ρ) = $3M / 4\pi r^3$

Where M = Mass of the object (kg)

r = Radius of the object (meters)

ρ = Density of the object

UNOFFICIAL COORDINATES, EXPLAINED

It is time to explain the unofficial seven-digit coordinates that were very briefly mentioned by Mr. Gaber at the beginning of the story. Gaber instructed Captain Travis to read a warning listed under the last coordinate on the list of destination coordinates found on the ship's monitor. The warning read as follows: Do not input any other seven-digit coordinate than those listed above; **the destination is not certain.**

It is a well-known fact that people often do what they are warned not to do. Therefore, I will briefly write down the unofficial coordinates and their destinations.

The unofficial seven-digit coordinates and their destinations are as follows:

0000001 - 0512373 Sun

0512375 - 1768123 Sun

1768125 - 2298445 Earth

2298448 - 3167207 Sun

3167210 - 3490074 Deimos, a satellite of Mars

3490076 - 4591334 Sun

4591336 - 4999999 Ceres, largest asteroid (asteroid belt)

5000000 - 5473888 Jupiter

5473890 - 5999999 Callisto, a satellite of Jupiter

6000000 - 6136089 Sun

6136091 - 6500000 Iapetus, a satellite of Saturn

6500001 - 6999999 Saturn

7000000 - 7500000 Oberon, a satellite of Uranus

7500001 - 7872038 Sun

7872040 - 8250000 Sun

8250001 - 8841425 Nereid, a satellite of Neptune

8841427 - 9296717 Comet Halley

9296719 - 9856020 Sun

9856022 - 9999998 The Milky Way's Edge

0000000 and 9999999 The ship self-destructs in five seconds

Made in the USA
Monee, IL
09 November 2026

67948862R00137